AIRLINE AVIONICS INSTITUTE

VOLARE AWARD

PRESENTED TO

MARIJAN JOZIC
KLM

FOR SIGNIFICANT INDIVIDUAL

OUTSTANDING ACHIEVEMENT

IN THE CATEGORY OF

AIRLINE AVIONICS MAINTENANCE

Awarded This Twenty Sixth Day Of April, 2004

Montreal, Canada

TALES OF THE CHAIRMAN

Hoofddorp 2017

The Netherlands

ISBN 978-0-359-33919-8

MARIJAN JOZIC

Marijan Jozic Development Leader KLM Royal Dutch Airlines.

Marijan was born in Croatia and earned most of his education there. He received his first Bachelors Degree from the University of Zagreb (Croatia) and become an Aerospace Technology Engineer. He received his second Bachelors Degree from the University of Amsterdam and became an Electronic and Telecommunication Engineer.

Marijan has worked for 40 years in the aviation business and served at KLM Royal Dutch Airlines in various positions: Flight Guidance Engineer, Systems Engineer, Project Manager, Business Analyst, Shop Engineering Manager, Development Manager, and Development Leader.

Marijan is a member of advisory committees at the Dutch Aerospace Laboratory and also at Avionics Magazine.

Besides his career at KLM, Marijan is also the Chairman of the Avionics Maintenance Conference (AMC) and a published author. Marijan has published several books about aviation and aerospace engineering activities and three books of short stories unrelated to aviation. He writes articles in three languages: English, Dutch, and Croatian, which are published in Plane Talk, Aviation Maintenance, Avionics Magazine, Data Bits, Matica Glina, PS-Portal, and Week van A&A. Some articles have even been translated and published in Russian and Turkish magazines.

He was nominated twice for the Aerospace Journalist Award (2004 and 2005).

Marijan received the Volare award in 2004, Total Quality Performance award in 2005, and Roger S. Goldberg award in 2017.

NOTE:

The book consists of articles, transcrips of speaches, motivational notes and interviews with AMC chairman in between 2012 and 2019.
Two articles inserted in the book I created together with Mr. Jim Saltigerald from Air Wisconsin and Kevin Kramer from US Air. Besides that there is also an article written about me by my dear friend John Lesly form Naasco.

The Tales of the Chairman

There was a traditional mid-term meeting of the AMC|AEEC leadership groups. For those who might not know, the AMC Steering Group and AEEC Executive Committee meet once between the annual AMC|AEEC conferences. Many things need to be arranged before the AMC|AEEC conference, and therefore, we meet in Autumn together. Each year, 20 to 25 aerospace engineers meet and do work that strives for a successful conference. This year there was no exception. If you look at the list of airlines attending the AMC|AEEC conferences, you will notice that many airlines located between Israel and Japan are not attending AMC|AEEC. Therefore, we decided to go to their area. The old saying is: if Mohammed is not going to the mountain, let's bring the mountain to Mohammed. In other words, let's go there and encourage their participation in AMC|AEEC. After careful consideration Bangkok was chosen. If you plan such an event, you must do a lot of preparation. It is not just arranging a hotel, but you must invite our buddies from the aviation engineering business and let them know that we are there for them. We have sent them an invitation letter and a package of data to show what we do and help them understand why we run such great Aviation Committees. Asia is vast; it is no wonder that Marco Polo spent more than a few years travelling through Asia on a horse. It is a huge distance, and we from the aviation industry take care to make possible that almost everyone can fly and get there in less than a day. So from all over the planet engineers were travelling to meet in Bangkok. Ok, back to the point. Our buddies from Thai Airways and Malaysia Airlines attended the meeting as invited guests. They all have great personalities, and they all are very enthusiastic engineers. It is funny that they also have the same problems and worries just like we do. It was great to talk and exchange information. The AMC|AEEC leadership committee members and guests from Asian airlines connected as if they had known each other for many years. There are no borders for aerospace engineering activity. They spent their time in the meetings and shared their experiences. It was just great. We decided to provide a program for our guests to brief them about our activities. It was amazing to hear the fellow engineers talking about the benefits of meeting and how AMC and AEEC changed their professional life. It was the AMC|AEEC at the best. But we also accomplished a lot of other work. The first day we met as joined AMC|AEEC committees, we discussed the upcoming Orlando conference and planned the joined AMC|AEEC symposium. For the joined symposium, we have chosen Aircraft Connectivity. I will not say more; I just want to make you excited about this e-enabled symposium. I am already very excited about the Orlando meeting.

The next day we split in two groups. The AEEC discussed new projects as well as adoptions of new standards or changes to the present standards. The AMC spent the day discussing two on-going projects (SCEA and Test Equivalency Working Groups) and also launching a new one: Design, Maintainability and Testability for Avionics Components. The AMC Steering Group has also chosen the symposiums for the Orlando conference. Yes, Intellectual Property is coming back again (just to let you know) and Mitch Klink will be the moderator. That seems to be a new AMC tradition. We reviewed the 2012 AMC survey results and tried to learn something from it. We were very pleased that we are getting great responses and high marks. That means that we are doing the right things. Some changes will be incorporated based on survey results. We have an obligation to keep the conference under constant improvement. We were also reminded how important it is to fill out the surveys to get as much feedback as possible. After completing the last few agenda items, we decided to call it a day. It was again great to meet and to do important things to keep the industry activity running. It was great to see our new friends, and we hope to see them in many future years at our AMC|AEEC conference. Just to let you know, I was in Bangkok, but **I have not** seen: The floating market, Laying Buda, Royal palace, Snake farm, Emerald Buda, Praya river, Lebua hotel with breath taking view from the restaurant at 64th floor, Pimman restaurant with Thai music and national dance, Rose garden, Thai boxing and other things. We did have a great meal in the Johnny Walker restaurant and ordered some great Thai food. Some food is very different than what we are used to eating, like Snake head fish, but due to a big storm we had to flee to prevent being flooded. As compensation the restaurant let us use their Karaoke setup. And let me tell you, the AMC and AEEC engineers can sing. No doubt!

Some say that the singing duet of Niblowich-Jozic compensated for all the flooding experiences. Although, some people said that they regret that they were not taken by the water. Their opinion was that it would be better than listening to Niblowich-Jozic singing.

Lower the costs. It is your choice!

Welcome to the second decade of the millennium. In perpetuity, it seems the last 110 years of aviation have faced challenges. Although the challenges have changed from basic flight to business, aviation is still an onerous industry. Some of the strategic challenges we face in this industry are immense. Intensive is a word that could describe air transport. Examples include capital expenditures, labor costs, customer service costs, operational costs, regulatory requirements. These are all intensive pressures on our business models. Consider another word: unpredictable. Fuel prices can (and do) change direction and magnitude as fast as the wind. We are extremely vulnerable to events out of our control. World events such as SARS, terrorism, volcanoes, and extreme weather are examples of the unpredictability that can (and do) affect our revenue streams. While we do our absolute best to maximize these revenue streams, one can not run a business without a focus on costs. We are all under extreme stress and in the end everybody complains that the costs are high and it is extremely difficult to maximize our margins to make a profit. Airline executives are always looking to reduce costs. Most of the time, fingers are pointing towards maintenance operations. Many technicians are doing their best to fix the aircraft or components. The modern air transport aircraft is very complex. A whole variety of problems has to be solved before our aircraft are fit for flight. Consider the following list as an example of the operational challenges we face daily.

Piece parts, Delivery of kits, Export licensing, Intellectual property, Test sets, Calibration, Equivalency, Contracting, Compatibility, Interchangeability, Intermixability, Obsolescence, PMA, DER repairs, COC's, DOA, Audits, Mandates, Airworthiness Directives, Maintenance programs, Heavy Mx checks, Dispatchability, AOM, AMM, CMM revisions, Software, Digital Mx records, Of course, this list is only the beginning. There are so many variables that it is better not to, know it all. You can sleep better if you are not aware of it., There are, however, many smart engineers who can see, the big picture. Sixty years ago, they were aware of complexity of aviation maintenance. They knew that something should be done to join forces and solve these challenges. Appropriately, they decided to establish the platform to meet, talk and solve daily problems. Granted, problems of the 1950's are different than problems of today. Therefore that platform is changing all the time. Now to reveal the secret. The platform is called AMC – Avionics Maintenance Conference. The whole purpose of AMC is to bring Airlines, OEM's and aircraft manufacturers together and solve problems. Every year more than 700 engineers join together. The best engineers from the airlines, the best engineers from OEM's and the best engineers from airframers join forces and solve hundreds of problems. Besides that, they do much more. Your question may be:

How does it operate? Engineers industry wide will submit discussion items and engineering questions. They actually describe their problem and submit it at the ARINC Industry Activities webpage. A note to consider for those who do not know: Industry Activities provides the structure to standardize the airline industry's engineering and maintenance issues, as well as organize and operate the AMC. Those submitted problems are collected and published - directly sent to airlines, OEM's and airframers. The idea is that smart engineers from industry look at it and come up with an idea to solve it. One person submits the problem and several hundred brains are activated. Once a year, the AMC takes place. This brain exercise results in solutions to the challenges. When a moderator calls the question, the airlines, OEM's and airframers will speak up with their comments, which results in bringing a solution to the problem. The airlines are happy, and the airframers and OEMs are proud to help their customer. It really is just that easy. And it is easy because there is a solid 60 years of tradition and structure in place. Every year, engineers return from the conference, happy and satisfied.

Just another success story.

Besides the AMC, the Airlines Electronic Engineering Committee (AEEC) meets separately at the same time and venue. This yields an additional 200+ engineers to the brain trust. Then there are symposiums, vendor hospitality suites, the Airline Avionics Institute reception and a lot of opportunities to network. Can you imagine that people voluntarily stay in a hotel for 4 days and talk avionics? They only talk, think, breath, drink and

eat avionics systems. Not just few individuals but hundreds of them. It is unbelievable. And why am I telling you this? Because I want you to know that there is a way to lower your costs in these difficult times. But it is your choice. You can ignore it and pay the highest price for maintenance on your aircraft or decide to join the avionics engineering community and be part of a joint solution.

The choice to lower your airlines costs is yours.

The next AMC conference is in Memphis. It will be the 62nd AMC, taking place between 18 and 21 April 2011. Again, we are talking about 62 years of tradition, value, and mutual benefit. This year FedEx is hosting the conference. Just attend the AMC in Memphis and you will certainly want to return in the future. You may even want to host the next conference in Anchorage 2012 or in Orlando in 2013.

Just register to attend the conference, click the link to reserve your room at the Marriott Memphis Downtown, and buy your round trip ticket to Memphis, Tennessee. It is that easy. In April 2011, Memphis will be the hottest place in universe. Don't miss it. Keep yourself informed via www.aviation-ia.com and, when the time comes, you will lower the costs of your maintenance. It is your choice. It is time to discover the AMC. You will not regret it. See you in Memphis.

What about the Rogue Units?

Written by: Marijan Jozic & Jim Saltigerald

If you search online for the term "rogue", you'll find one of the following responses: *"An organism that shows an undesirable variation from a standard. An inferior or defective specimen, or a wicked or evil person; someone who does evil deliberately."* I prefer the latter definition: *"evil."* That is an accurate definition when applied to troublesome aviation parts: something evil that is constantly traveling between the airplane and the avionics shop. Complete elimination of the rogue unit phenomenon is not possible, but it can be managed. The good news is rogue units can be managed. The bad news is that they cannot not be banished, and managing the units comes at substantial costs. While the definition applies well to aviation, there is an industry definition that avionics engineers prefer. To begin with there are three levels for troublesome units: 1st tier definition - Rogue Unit 2nd tier definition - Chronic Unit 3rd tier definition - Poor Performer **Avionics Engineering Rogue Unit Definition (1st tier)** 1) Unit has a history (3 unscheduled removals) of repeated short service. 2) Demonstrates repeated identical system faults, i.e., the same complaint every time. 3) A replacement component of the same type resolves the aircraft system fault. 4) Standard bench or overhaul testing procedures have not detected the cause of failure. I've had personal experience with the last qualifier. I recall a generator control unit, which failed after we accomplished the end-to-end test for 57th time. It was reminiscent of those AMC questions when an airline asks, "Is it safe to recertify the LRU if you run the test 5 times and it failed the first time and passed 4 times after that?" Or, "What if it passes testing every time and comes back after the first flight after installation?".

Chronic Unit definition (2nd tier)

1) It has a history (3 unscheduled removals) of repeated short service. Repeated short service has to be agreed.

2) Demonstrates repeated system faults which may or may not always be the same complaint.

3) A replacement component of the same type resolves the aircraft system fault.

4) The cause of failure has or has not been detected by standard bench or overhaul testing procedures. These units are particularly frustrating. Each time maintenance personnel removes them from the aircraft due to a failure, something different needs replacing. The definitions above clearly define the difference between the Rogue and the Chronic unit from an airline's perspective. The airline, in most cases, is in the best position to make the initial determination between a rogue and chronic unit.

However, vendors are also taking a proactive stance and have a process in place to help mitigate risk. It is important to emphasize that both airline avionic shops and vendors are interested in solving the problem. Avionics shops are interested because they want those units flying and not hanging in the shop. Vendors are concerned because they know their reputation will suffer if engineers tell each other about troublesome units and their respective vendors. There appears to be a few scenarios that play out in tracking and tracing rogue unit component performance from company to company. Just to name a few

- Detection & Discovery Scenario 1: Airline to/from OEM

- Detection & Discovery Scenario 2: Airline to/from MRO
- Detection & Discovery Scenario 3: Airline to/from Third-party
- Detection & Discovery Scenario 4: OEM to/from MRO
- Detection & Discovery Scenario 5: MRO to/from Third-party
- Detection & Discovery Scenario 6: Third-party to/from OEM

Of course, the above scenarios could become a triangular conversation, and the basics still apply in managing through resolution and closure at the local level. In all the scenarios described above, the resolution begins upon detection (data points to a suspected condition) and discovery (data is verified to indicate that the condition does exist) of very poor component performance. This then becomes a local issue, managed between the two parties involved (in this case between Airline and MRO). If the Rogue Unit Policy applies, then it must remain between the parties dealing with detection and discovery. For example, an airline discovers a rogue unit according to the criteria of the definition. If the airline has an agreement with an MRO, the main points of contact in working towards resolution and closure should be between the airline and the MRO. Should anyone else need to be involved, those parties should provide a supporting role. Next is the trickier part of dealing with rogue units. There will be instances where the performance of a rogue unit will not meet its reliability target. In this case, tracking is needed to prevent it from becoming a "hidden failure". In today's current applied technology, when dealing with discovered rogue units, the airline operator is better able to directly address the issue with the original equipment manufacturer or vendor. If all remedial actions have been exhausted by the parties involved and if there is no sign of improvement, the unit should be exchanged or discarded and replaced. Purchasing a new unit can be more cost effective than tolerating failures for many years. Just a few notes about that: My personal experience was a problem with the Air Data Computer installed in one of our fleet of single-aisle aircraft. There was an intermittent failure in one of the serial numbers (a single ADC LRU). The ADC was flying just a few flights and coming back to shop with a complaint: *Airspeed flag intermittently in view during the cruise*. After a few months of problems somebody said: "why don't we throw it in the lake nearby and buy a new one"? Due to the $30,000 USD cost of a new unit, management turned down the suggestion. Well, it is now 20 years later. The ADC was removed more than 250 times from the airplane type. This roughly averages 12 times per year, i.e., once a month. If the average shop visit costs $300 USD, the estimated cost of service for this part was $75,000 USD, excluding the invisible costs of different investigations, meetings, fax writing (when it started there was no e-mail) and later e-mails, removal and installations at the flight line, low range leak testing after every installation, and a lot of frustration. After 20 years and a long series of discussions, the unit was sold to a broker for $1,000 USD, and we thought we saw the last of it. After another long series of discussions, somebody from a totally different department decided not to buy a new ADC as a replacement but "a good used serviceable ADC with fresh serviceable certificate". And, you can see this coming, that same ADC came back in the inventory for lousy $3,000 USD.? Therefore, a word of caution: a rogue may change by part number, or even by dash number. It may return into serviceable inventory, but the bottom line is that it becomes a local issue no matter where it resides and how it is detected. Rogue units should be managed locally and at each iteration as such. In a situation whereby an airline has a component agreement with an MRO and the MRO keeps a database of all available component removals, it could be the MRO that addresses the issue on behalf of the airline operator directly with the original equipment manufacturer/vendor or other third-party. When known performance deterioration is detected, to the point where discovery is met, then the issue needs to be addressed as suggested by the Rogue Unit Policy definition.

The rogue unit definition is a starting reference (alert indicator) point from where to launch an investigation. Once identified and remedial action is undertaken, the options should be to:

a) Exchange the unit for a known good unit (preferably accompanied with teardown history).

b) Receive the unit back and expeditiously perform a control-fit and continue to work through resolution.

c) Return the rogue unit and "tag" the rogue unit as a "hidden" failure by fleet model type, part number, and serial number, and submit a discussion item for the next AMC and share your lessons learned. This has to be the starting point of visibility and the "local" level referred to for uncovering "hidden failure" rogues. It can then become a discussion item at the next AMC. Usually what worked/didn't work can be shared, and other Airlines/Vendors/MRO/Third-parties at the AMC can learn from their experience. We could shed some light on the issue by giving these units visibility in an open forum. At this time, I can only think of the AMC as a place to flag units after all remedial action to improve performance has been exhausted and unsuccessful, and, where it could be done neutrally. Since this topic involves Supplier Support Agreements, it has been recommended that this issue can be added to the ARINC Standard for Cost Effective Acquisition (SCEA) Working Group. On the topic of financial impact, please take a moment to browse the link below. Everyone is encouraged to participate in this activity. http://www.aviation-ia.com/amc/projects/index.html Then click on *Standard for Cost Effective Acquisition (SCEA) Working Group* A final note: We all know the subject has not been thoroughly exhausted so here's your chance to voice your opinions and solutions. The SCEA Working Group is your opportunity.

Can you see the invisible?

I recently found a saying on an the internet: "When You See the Invisible, You Can Do the Impossible." - *Oral Roberts* Many times each of us avionics engineers realizes that it is good to see invisible; that there is something going on which is causing problems. After you see it, there is one other strong feeling which commands your attention. You get the urge to define the problem very clearly and make it visible to others. The next logical step is to solve it. Some will say that it is impossible, but you know the best way forward. Avionics engineers are able to access the data in the aircraft maintenance system. It can be the data about reliability, shop findings, costs, design – you name it. If an avionics engineer takes some time to analyze, he will immediately find answers in some of those parameters. Let's start with reliability: there is a lot of data. Every airline must collect the data, but there is no mandate to analyze it. Analyzing the reliability data can give you an idea about the performance of your LRUs, but there is a catch. Some airlines choose to depart on time. Due to that policy, the first line technicians, suffering from a lack of troubleshooting time, choose to replace a suspected LRU. Sometimes, it may be even two or three LRUs. That is called the Shotgun troubleshooting technique: just keep replacing boxes until the system starts to work properly. In shotgun maintenance, the first line creates NFFs (No Fault Found). This is the choice of the company, and as the saying goes, that's the price that goes along with it. Of course, no one likes NFF but nobody likes delays either. The big question is what is more expensive: NFF or delay? The problem is that some airlines have an NFF average of 40% or more. Explained another way, this means that 40% of component removals are actually good LRUs. There are two ways to look at it. One way is comparing the cost of one NFF against one delay. Some airlines have a fixed price of delay. For example, the unplanned hour delay of a B747-400 type of aircraft can cost $100,000 USD. This includes costs of crew, gate position, missed connecting flight, etc. On the other hand, test and certification of an FMC (Flight Management Computer), for example, would cost not more than $1,000 USD. So, in this case the decision to replace the

FMC and MCDU would be acceptable. The MCDU is failing, but the technician chose to replace both LRUs, eventually finding one box faulty and one fully functional. In the end, the flight departed on time and we only have one NFF. In this respect, it was cheaper to have a NFF. But if you have 100 removals per year and 40% NFF, this means that you spend $40,000 USD every year removing good LRUs. That is quite expensive.

Of course, the avionics engineer should do something about this, because he sees the invisible problem. Although everybody is happy about an on time departure, the same people would not be pleased to learn that they are spending $40K for removing good LRUs. That $40K is plenty of money to organize a short refresher OJT class for technicians to teach them how to troubleshoot and recognize the fault(s), and remove only failed LRUs. Let them learn to see the invisible. Shop findings are another issue which can help you to see the invisible. First, you must teach your technicians the discipline to fill out the shop findings form properly and note the relevant information. It is vitally important to document the LRU's condition and the failure symptoms. You have probably seen the complaint: "Failed" and shop finding: "Repaired". The next time you see that, go to the technician and beat him with a thick stick. It is in his best interest to document the system's failure and describe what is found and what is replaced. Note which part of the test failed and what was measured (e.g., voltage, etc.). That will assist the technician (and you) in the future. If you do nothing, the problem will remain invisible. One day you may have an accident, and there will be an investigation. The end of the story is that you will have to solve the root cause of the problem. For many people the problem was invisible, but you knew about it and did nothing. Can you sleep well at night when you recognize that there is a problem and you are doing nothing? By monitoring your LRU conditions, you will notice that some parts are in the shop more than other parts. There is an increase in the costs of repair of those LRUs. Since you have access to data, investigate and figure out what is causing the problem. At this stage, it is not important which LRU it may be, but it is important that you saw the invisible. You recognized that the costs were related to certain PCBs (Printed Circuit Board). The shop is replacing the PCBs and simply throwing the old ones away. If you go one step further, you will notice that there was no P/N of the piece part in the CMM, and they were ordering the next higher assembly and not documenting that in the maintenance logs. You can not blame them because they are following the CMM: that is what they do. Well, if you negotiate the addition of the P/N of the part into CMM, and I admit that sometimes it is not easy, you can save valuable time and money for your shop. In this case, the P/N of the piece part was invisible in the CMM (and to your technicians) but not for you. So if you can see the invisible, you can sometimes accomplish a seemingly impossible task. Some time ago, I asked a technician why they were throwing so many of those square coffee pots. "Because they are leaking," he said. "I am avionics guy, but I can see that it is not impossible to weld those holes and prevent leaks," I responded. To prove this article true, he continued, "Since there is no procedure in the CMM, we can not do it."Well the invisible procedure was in his head and in my head too. But now we just have to legalize it by defining the instructions; checking the feasibility; and after discussing it with the item's engineer and your Designated Engineering Representative (DER), it is fixed in no time. No more leaks and saving $2,000 USD per event. The conclusion: the savings are there, but you should see it in the first place. If it stays invisible, it will never generate the savings and you will not be able to make the impossible. Managers often urge people to define their processes, which (in theory) can make all those things visible. Managers are *always* talking about processes. That is a fact, but that is exactly the problem. You can not just define a process flow to discover potential savings, and then based on that discovery, generate procedure steps to make the savings visible. In avionics, every time it is something different and new. Every time it is in different area. Sometimes it is in the procedure, sometimes in CMM. The same is true for the drawings, technology, company policy, and sometimes simply in sound engineering judgment. But our true power is in the work force. The biggest gains are made if the work force is trained to see the invisible. Only then can they accomplish the impossible. Sometimes it is not easy, but that is what makes our avionics life so exciting!

Notes from the Chairman 1.0

The news is traveling fast; but for people who are not aware, there is a change in the AMC Steering Group. Mitch Klink, FedEx, has decided to step down as Chairman. It is time to say: Thank you Mitch for your leadership during last three years. This is not an easy job, and we were very happy to have you in the position. It was our pleasure and privilege to work with you. I will now deliver the next news: The new AMC Chairman becomes Marijan Jozic (KLM). Leading AMC is a big challenge.

The aviation industry is stressful, and operators are moving from one crisis into another. Everybody is under pressure. For the last 3 years under Mitch's leadership, the steering group has been working to benefit the industry in different areas. Let's touch base with some of them. High on the agenda was the quality of AMC conference. The AMC attendees are requested to complete the yellow forms (surveys). The AMC Steering Group reviews the surveys so that we can measure important parameters. The voice of our customers (attendees) is always an important part of our discussions. This is especially true of your suggestions about improvements or original new developments in the industry of interest to us all. Each time there were many good ideas but, alas, not enough resources to implement them all. Special attention is given to symposiums topics.

The symposiums are to be educational, but they should also address important issues. At the top of the list every year has been the Intellectual Property (IP) symposium. This was so important that in Anchorage this year we decided to continue the discussion started at last year's panel symposium. People who attended both seminars discovered that in just one year a big change in awareness was created. This is the added value of AMC. I am sorry for the operators that do not attend AMC; they will have a hard time catching up. In 2009, the AMC and AEEC collocated. The two meetings in one location is very cohesive. Both groups of engineers became friends and joined forces in developing cool new equipment and maintaining the old equipment. It is great to see that the conferences became better than everybody expected.

Mitch Klink, AMC Chairman (2009-2012), is a person of few words. But when he says something, you better listen. He is very knowledgeable, and after short talk, you notice that he is extremely experienced. Just the type of engineer we all want to be. He is the true example of who every avionics engineer should strive to become. We are happy that Mitch is still around and that he will keep supporting the AMC. For now we will say THANK YOU Mitch for 3 years of chairmanship and thank you in advance for your continued support as the AMC's new Vice Chairman. And now something about the future: the next AMC will be in Orlando, Florida.

But before Orlando, there is a lot of work for all of us. There are some items that should be handled. Those matters are applicable to all of us, not only for the new chairman. In the first place, we all know the AMC formula –Identifying hot issues that not only concern your organization, but probably the industry as a whole. Opening the dialogue about these discussion items is what the AMC is all about. That will not change. So in preparation for the 2013 AMC in Orlando, review these ideas: Start collecting questions right now. Make a folder in your computer, and every time you have a problem, note it in the folder. Notify it in the same format we use at the AMC and it will save you some time later. Try to estimate the cost savings. That number is important for us on the steering group to calculate the savings for the whole industry, and it will be important for you to show your management the benefit of attending the AMC. Promote the spirit of AMC everywhere. Tell people that we are producing the ARINC Standards that promote efficiency in our industry. If you meet people of an airline who have never attended the AMC, tell them what it is about. Tell them that we need them and that they should also become an AMC Member Organization. That will give them access to a great deal of useful data as well as being a part of the biggest avionics network they can imagine.

Many Asian airlines are completely isolated from the aviation community. I have personally witnessed that. When you talk to them, they have that urge to be helped. If you speak to them, just tell about the AMC power and sooner or later they will be a part of our family. The AMC Steering Group will have two meetings before the next AMC. One in the fall of 2012 where we traditionally review the surveys and decide what changes to make for the next conference. Also, the ARINC Standards activities are discussed, and the decision is made which documents should be updated and which new projects will be initiated. Every steering member provides the information about his or her situation in his or her part of the world. Additionally, symposium topics will be reviewed and determined. Usually we will have 20 or more submissions for symposiums, and the AMC Steering Group will determine which of them will benefit the AMC the most. After a determination is made on symposium topics, a moderator is assigned. Those individuals take the responsibility to contact the speakers and arrange all necessary details. The audience will only see that the symposium was smooth and interesting. All the work behind the scenes is invisible. Therefore thank you moderators for all the hard work.

The second steering meeting is traditionally in January. That is the final preparation meeting before the next AMC. It is an important meeting because it is where we review the discussion item list for the open forum. Before end of the steering group meeting, the detailed planning for the AMC is reviewed. The roles and responsibilities are defined and the final agenda is settled. The next time when we meet, we will be at the hottest place in the universe: the AMC. It is great to know that we have such great tradition and that we have been holding the conference for more than 60 years. It does not matter where we meet, it does not matter when we meet, but we all know that the most important week in the avionics world is the week of AMC. Not only for the steering group but all of the airframers, OEM's, MRO's, airlines, and operators are meeting that week. But there are challenges. The aviation industry is in a state of perpetual upheaval. It looks like turmoil is standard operation nowadays. It is now more than ever that we need each other. I am not talking only about operators but about airframers and suppliers as well. It will be good that everyone starts to realize that the ultimate goal is to sell safe, efficient travel in our airplane. We have worked hard to increase the level of safety so high that we hardly hear about accidents and incidents.

Our passengers are not looking for the airline with highest safety record. They are looking for the airline offering inexpensive seats. The lower the seat cost the more likely consumers will purchase. It is just basic economics. Unfortunately, we have spoiled the passengers. They would like to pay less for flights on our airplane than they pay for a bus ride or train heading in the same direction (well, except crossing the ocean). This is of course the paradox of our time.

To keep our prices low and be competitive, the airlines need to purchase new airplanes. The airlines need to reduce operational expenses. Airframers want to help the industry by selling new airplanes that have low fuel consumption. Well, the whole chain is complicated, but the bottom line is that if operators make money, they will buy the airplane, LRUs, modification, or maintenance services. If operators of the aircraft cannot generate cash, they will not buy anything from suppliers and the whole logistics chain will stop. For the past few years, we have been balancing at that edge. This is not good. Therefore, the next couple of years will be important. We all will have to adapt to the rules of a new game. At this moment, it is not easy to predict what will happen in next 12 months, but the AMC will keep you posted. By the way, you can also keep us posted.

AMC has survived 63 years because of a unique formula. That does not necessary mean that the formula will not evolve. We older guys notice that the types of questions are different. Our first two days of recent years' AMC we are discussing maintenance philosophy and product support. Twenty-five years ago, we did not have questions about maintenance philosophy. Change is good, but to stay in business, you must change in the right direction. Therefore, we will have a lot of work to do. We will have to anticipate and make subtle changes because the whole world is changing. We must adapt to new situations or else we finish up as dinosaurs. As the Chairman of AMC, I will promise that together with the AMC Steering Group I will do everything to continue the strong leadership of the AMC.

What is in Common?

What is in common between a Ford Model T and a Tesla Roadster? First let me tell you something about the Ford Model T (known as the Tin Lizzie). The Model T is a car that was produced by Henry Ford's Ford Motor Company from September 1908 to October 1927. It is generally regarded as the first affordable automobile, the car that opened travel to the common middle-class American. The Model T made 1908 the historic year that the automobile became popular. One could purchase a Model T "in any color so long as it is black." Tesla Motors is named after electrical engineer and physicist Nikola Tesla. (Note: Nikola Tesla was born not more than 60 miles from my birth place in Croatia.)

The Tesla Roadster uses an AC motor descended directly from Tesla's original 1882 design. The Tesla Roadster, the company's first vehicle (2008), is the first production automobile to use lithium-ion battery cells and the first production electric vehicle with a range greater than 200 miles (320 km) per charge. If somebody asked me what is the commonality between those two cars with respect to maintenance, I would think that he is not really smart. It would be obvious that he is not able to see the difference. For this hypothetical, inquisitive person, there is obviously no difference between a Ford T and a Tesla. He is not able to see the difference in brakes, steering wheel, windshield, sensors, indicators, seats, doors, etc. Besides that they are vehicles built for the transportation of people on the road, the only commonality is that both can be delivered in black. The question is, where this story is going to.

The Boeing 787 is flying and more are being delivered. Soon hundreds of them will be flying. The old story says that everything that engineers build will fail one day. Well, it inevitably will happen with parts installed in the B787. If you look at statistics, it will take some time before the first components will fail. It is amazing how good the components are today. If you look at the lists of LRU's, there will be two numbers which are large. In the first place the costs of LRUs (colossal), and in the second place the reliability (huge). In the 1970's and '80s, we were happy if an LRU would be able to fly for 4,000 hours. Later on when the B747-400 started to fly, the numbers were increased to 15,000 or more. And for the B777 and A330, the parts are flying 30,000 hours or more. If the numbers are correct, we can expect that the B787 components (and later on the Airbus A350's) will reach 50,000 hours or more. Now that's talking about a great MTBUR. How big is the number 50,000 hours? Well, it is a scary number, frightening for avionics shops. Can you imagine that an average B787 flies 4,000 hours per year (let's assume that the number is correct; not far from average). The simple calculation will show that we will have to wait 12 years to get the first component in the shop. If we extrapolate it to a fleet of 25 aircraft, that means that there will be just 2 removals per year (assuming that there is one LRU per aircraft). For two LRU's per aircraft, we will have 4 removals per year. That is actually a dour prediction for average avionics shops. My calculation might be a bit off, but let's pretend that is in that order. That means that for certain LRUs, your test set will be switched on 4 times a year. The new LRUs are complicated, but the test sets are complicated too. Test sets are able to check the LRU and almost pin point where a failure is. The repair level 2 is easy, just swap the printed circuit board and go. Per LRU, there will be not more that 2 or 3 hours effort including the paperwork. With 4 LRU's per year and 3 hours of technicians work, you can't maintain that part economically. There is just no way. After the above description, we came to a dilemma. Is it worth it to purchase extremely expensive TPS, train our people, keep a stock of spare parts, and keep the test set within calibration all just for 4 shop visits of 3 hours each? That is

12 hours a year. The above simple story tells me that it will be that very rare for shops to invest a lot of cash for test sets. It is not reasonable to invest for just few shop visits. Therefore, there is a big chance that we will have just a few big shops in the world which will do the B787 and A350 components. Most of the B787 operators will decide to outsource the component repair. It is just not economical to keep the avionics shop for B787 components if the removal rate is low. The folks buying only a few B787s or A350s are not stupid. They know how to calculate the return on investment. On another side, you need a big fleet under contract to be able to feed the shop with broken components. To get the decent flow, say 40 components of one kind per year, you would need at least 200 aircraft under contract. That is a lot. But there is no other way. If there are not 40 components of a certain kind per year, there is no way to finance the test set (or TPS). There is a new big problem. That is the comparison of costs of maintenance of those components with respect to previous generation of the components. At this moment, nobody has the experience with the components because there are no removals. Also, the components are still under warranty. So, you even can't benchmark it against the price you pay to the OEM for repair. There are several approaches to get the average repair price. One of them is to compare the B787 component with lookalike components installed in B747 or in another legacy aircraft. The financial guys will shamelessly send the list of B787 components to the engineers and ask for them to find the lookalike component which is used in old aircraft type and ask to do the comparison. It is like comparison between Ford Model T and Tesla. Yes, they both have brakes, but can you compare the costs of maintenance of brakes of a Model-T and Tesla (mechanical vs. electrical). How can you compare the maintenance costs of a common core system in a B787 with the flight management system on a B747. They have the lookalike functions, but there is totally different philosophy imbedded in the concept. There is an engine in the Model T and an engine in the Tesla. Are they comparable? The comparison method I described is lousy and totally inaccurate. But, the financial guys need something to calculate and they also need somebody to blame later if the calculations are wrong. They also don't take no for an answer. If you tell them that there are no similar parts for liquid cooling, head up displays, fiber

optic converters, data concentrators, electric actuators, or solid state circuit breakers, they will not believe it. Little by little, we are stuck in the situation where the gap between engineers and non-engineers is getting bigger and bigger. They speak different languages, and with new technologies the difference between languages even bigger. The question is who can provide the dictionary to translate engineering language to commercial language? Well, the answer is obvious: the engineers. (Oh boy, it is again the same story: everybody hates engineers but nobody can live without them.) The financial guys seemingly have not evolved. They still use the same spreadsheets as 20 years ago, and they can still produce the most wonderful lists just as many years ago. But if the philosophy is "Garbage in, garbage out", they will never ever have good maintenance cost estimate and they will never be able to provide the reliable offer to their customer. Meanwhile, engineers were learning all the time, because the engineering environment is changing all the time. They have actually done the terrific job making the components so reliable that engineers become redundant. There is such a huge technology difference created that you can't compare old components with new ones. The prediction of maintenance costs for next few years will be just a guess until we gain sufficient data and gain sufficient experience. We all have to experience the new generation of aircraft. (I almost said aluminum birds, but I realized that they are not aluminum any more.) We all know that the average repair of a new component will not take many hours. Yes, the components are complicated, but there is very sophisticated built-in-tests and the failure diagnosis will be faster (at least I expect that). Once the failure is isolated, the part will be changed and the test will follow. Many of the components will be repaired very efficient as long as you adhere to level 2 repairs. But if you go to level 3, it is a new ball game. Some components have multi-layer circuit boards. Some OEM's are telling us that even they are not repairing multi-layers boards, especially if there are 10-15 or more layers. Again the same story! They don't want to invest a mountain of money for expensive equipment to diagnose failures if they only get 2 or 3 broken boards per year. Repairing multi-layer boards is a great challenge. Again, there is a translation problem to explain it to the commercial guys! The lead time of repair is not the time required to unsolder and solder a capacitor on the board. It is the time you spend to figure out the failure at the board level. Sometimes it takes the whole day to find the problem and 5 minutes to fix it. The customer is willing to pay 5 minutes but not the 8 hours. Why? Because he is not aware of how complex the electronics are. He is still on Ford model T level. We did not even discuss software. Well in a Ford model T there was no software. In a Tesla Roadster, there is a lot of software. I don't even have to check with the manufacturer, because it is just a fact of life. When we started B747-classic operations, there was no loadable software.

Later on in B747-400s we started with 10 loadable systems. It extrapolates to a few hundred loadable systems on B777. Every piece of today's technology has software inside. You can ask yourself why the lavatory flush control unit needs microprocessor controls. But it does, and it is probably more reliable and easy to manufacture. The question is if the technicians that previously worked on mechanical parts will be able to switch to digital systems and all electrical components. Alas, the B787 is one big flying computer system. There are computers everywhere. The need for electrical energy is so huge that there are two generators on each engine. And it seems that every single part has a microprocessor. All that computer stuff is using a lot of energy, and it can't be cooled with air. That is the reason for liquid cooling. Something not seen in present type of aircraft. There are many reasons that I don't like the comparison between the B787 and the B747. It is not helpful. The comparison will produce a list of unreliable numbers. The only good way to do it is to give engineers enough time to analyze the CMM's and calculate costs for repair for each LRU. It will take some time, but then at least you start with reliable data. Now is the time for a friendly warning for all airlines that may be potential B787 and A350 operators. Start preparations sooner than later, and be aware that now is the time to work on your contracts. Use at least the SCEA document (ARINC Report 674) to make sure that your contracts are acceptable. They should cover the life cycle of your aircraft. The comparison in technology between B787 and old legacy aircraft is not the good way to go. Comparisons between contracts for the B747 and B787 are not good either. The B787 contracts must be much better, and don't even try to compare it with old ones. Many years ago when you wanted to buy a brand new DC-3, there was just one page of paper. A B787 or A350 contract is almost such big pile of paper that the average engineer can't lift it. There is one new important point required for B787 contract which can't be found in old contracts. It is the Intellectual Property (IP) chapter. That will be the name of the game in next 10 years or so. We all know that the IP battles started two or three years ago. The comparisons between B747 and 787 or between a Model-T and a Tesla Roadster are useless also in the area of contracts. But that is a whole new story. Back to the comparison between the Ford Model T and the Tesla? Which car would you like to drive to your office? I know my answer.

Gables Engineering!

"Of course, it's Tradition!"

What does a Croatian watchmaker, the world oldest airline, and a manufacturer of control panels have in common? The tradition!

Let me explain.

In 1892 my great grandfather decided to stop repairing church clocks in Austria because he got too old to climb in the church towers. He opened a watchmaker's shop in the Austrian-Hungarian Monarchy. The shop was passed from generation to generation and is still doing repairs of the watches. It is a long family tradition.

The oldest airline using the same name is KLM. KLM was founded on October 7, 1919 by Albert Plesman, making it the oldest carrier in the world still operating under its original name, though the company stopped operating during the Second World War—apart from the operations in the Dutch Antilles in the Caribbean. The first KLM flight was on May 17, 1920, from Croydon Airport, London to Amsterdam carrying two British journalists and a number of newspapers. Today, KLM still operates under the same name. It is a long tradition.

In Coral Gables in 1946, Mr. Victor P. Clarke began making control panels. Dedicated to offering a superior product, Mr. Clarke founded Gables Engineering by combining a keen understanding of manufacturing techniques with a superior knowledge of aviation electronics and design. Gables is the only vendor still operating under its original name and as an independent company. Again, it is a long tradition.

Combined with a thorough knowledge of the aviation industry, Gables rapidly established an international reputation as the source for reliable custom-engineered control panels, audio systems, and related products. So now picture this: the AMC Chairman, descendant of a Croatian watchmaker with a long tradition of repairing watches, presently working for KLM, the oldest airline in the world, visiting Gables Engineering in January 2013. It is probably a coincidence but let me conclude that there is a tradition involved.

Tradition means everything. Today, under the leadership of Mr. Victor E. Clarke, Gables Engineering Inc. remains the industry leader in custom avionics controls and other related products. Gables' engineers continue the tradition today by maintaining a sharp focus on all aspects of control panel technology as it exists today and as it continues to evolve. That is exactly the point. If you want to keep the tradition going you must evolve. That is the only way to remain in the aviation business. Today, it seems times are hard. The times were always hard and the times will continue to be hard. Without evolving and changing you can't play a leading role and keep the tradition in place. I must admit that my path has crossed Gables' path several times and it was always something to remember. I started as control panel repairman at KLM in in 1980. The very first day at work I got a VHF-NAV control panel in my hand. That was the one with one knob at the front and a mechanical dial. Inside was a long mechanical multi-deck switch (2 out of 5). I don't remember the p/n but it started with G (what else).

And now the funny coincidence!

When I visited Gables in January 2013, I parked a car behind the building, locked it and approached the building. In front of the building there was a nice gentleman and I asked him where the main entrance was and was it in fact Gables Engineering. You can't trust GPS in the car. The GPS told me: "You have arrived!" But there was no sign indicating Gables Engineering. He showed me the entrance, smiled, and said, "We don't need a sign. Everybody knows Gables Engineering!" Later on we were discussing the traditions and I mentioned the VHF-NAV control panel as my first job. One of the younger Gables engineers asked me if I remembered the guy from the parking lot. He was the engineer who designed the VHF-NAV control panel I repaired back in 1980. WOW! That is absolutely fantastic. That is passion! That is dedication! That is aviation! That is love!

Back to the second part of the 1980's. It was the time when modern aviation started. According to my definition it was kicked off with TCAS.

KLM wanted something different. The certified control panel per the STC was not good enough. The pilots wanted a panel with a keyboard. The only manufacturer who was willing and able to build it for us was Gables. The ATC/TCAS control panel with the keyboard and small LCD display was introduced in the old fashioned, 747-jurassic cockpit. Gables helped us again, although we first specified a gray panel with black buttons and then changed it to gray panel with light gray buttons. Just a simple engineering and p/n change. Not many years later (in 1995 or 1996), there was that project about 8.33 KHz channel spacing and mandate for FM immunity. A Gables panel for VHF-COM was installed and because the VHF-NAV was involved we decided to install a FM Immune MMR and do the certification at the end of the year. With no time to spare, there was a crisis in the project. The EMI test in lab in California had a problem and we could not explain why. It was Christmas weekend almost no chance that we could get somebody to help us. But Gables Engineering did not hesitate a second: "Yes, we will send an engineer to Fullerton California." And they did. The Gables engineer offered his Christmas holiday to help the customer. The SB was issued in short order and the project was saved. Isn't that passion for aviation? Isn't that dedication? Good job, Gables, I will never forget it! And now, just a few words about the great people at Gables Engineering. The AMC Steering Committee meeting took place at the Gables facility in January 2013. As you know, Gables is hosting the 64th AMC and it was a good opportunity to see them and fine tune the program. Well, the Steering Committee consists of experienced professionals and most of them have seen many facilities but Gables Engineering was something different. It's a great place, clean and spotless and with lot of nice smiling people. They have continually invested in their plant and equipment, and have developed many proprietary processes that give them an edge in the market. They also think about employees. There is a full blown gym on Gables' property and it is well attended and used by employees. Top management as well as all other employees can enjoy playing a basketball game or lifting weights together. Due to sport, the great atmosphere is created and that all has an effect on team spirit. Due to daily exercise, people are healthier and cheerful and happier. Believe me, happy people produce quality, and that is Gables. As it was in 1946, they continue to provide their customers with excellent, well designed and timely executed custom solutions. Gables currently maintains an active LRU catalogue of hundreds of part numbers, over 100,000 examples of which today circle the globe with virtually every major operator in existence on virtually every commercial aircraft produced over the last 67 years. This is just my short impression about Gables Engineering, host of the 64th AMC in Orlando. Not to mention all the great AMCs that we shared fun in Gables hospitality suites as well as in the big ballroom. Speaking about open forum discussion, there were not many questions for Gables engineering and for few of those you could hear the moderator announcing: "This is one more AMC success story."

Gables! "Of course. It's tradition."

Chairman's Welcome Orlando Florida

Article in the AMC program for Orlando conference 2013

On behalf of the AMC Steering Group, we look forward to seeing you for the 64th annual AMC|AEEC in Lake Buena Vista, April 22-25, 2013, hosted by Gables Engineering. We sincerely appreciate the continued support from our Member Organizations and Corporate Sponsors of ARINC Industry Activities that enables us to conduct this one of a kind global event for the benefit of all our attendees. As an Operator attending AMC, please spread the word about who we are and what we do. As a Supplier or Airframer, please remember that being an AAI member is an important contribution to AMC, but if your organization has experienced increasing success as a result of participating in AMC, please consider also becoming a Corporate Sponsor of ARINC Industry Activities.

By growing our Memberships and Sponsorships, we can ensure the future of AMC as your primary source for constructively resolving technical issues, providing information exchange, and networking with business partners you might only see this time of year. The AMC is unique to our Industry as the only technical forum that brings the Operators, Suppliers, and Airframers together in one place to work towards reducing the lifecycle costs associated with maintaining avionics. Please remember, avionics is anything with a wire on it! Ensuring we provide safe and reliable air transport in a cost effective manner is of upmost importance for our Industry's survival. This involves managing the life-cycles of our aircraft and components beyond what we normally consider to be traditional maintenance – warranty, repair order issues, timely delivery of piece parts, etc., are all within the realm of what AMC can assist you with. While we have had some bright spots in our Industry during the last year, there are many Operators that are still facing serious economic challenges as we forge ahead.

As an Airframer or Supplier, your business philosophy should be to help the Operators be successful. Your success depends on the success of the Airlines – we are your customers, not your competitors. Even though the primary goal of AMC is to resolve technical issues, we cannot ignore the commercial aspects of our business as each of the key players struggle for their slice of the pie. Only through the information exchange possible at AMC can each of these key players understand the important aspects of each other's business, and hopefully, respect the boundaries necessary to assist your partners achieve their goals and objectives.

Proudly serving, Marijan Jozic

In Memoriam Mr –T (Japan Airlines)

If you have not heard, I'm sad to report that Nagashige Toritamari of Japan Airlines and a long standing AMC Steering Group member passed away this month. Mr. Nagashige Toritamari was known as Mr. T. was a jewel among us, always giving us a shining example of kindness, caring, and fun. Although distance and time took us in different directions through life, he was a firm believer in the mission of AMC, as well as a close and great friend to all AMC people. The last time the AMC or the AMC Steering Group saw Mr. T. was during last year's AMC conference in Minneapolis. Even though he had just flown some 20 hours to attend the AMC event, he arrived with a smile on his face. He was a man of few words; however, when he spoke, he always had something meaningful to say. Many did not know, but Mr.T. was fluent in German, and you could occasionally find him in the Gables Suite singing the famous German love song *Lili Marleen*. It was our privilege to serve the AMC together with him and to be his fellow engineer and friend.

May God bless his family during this time of sorrow.

AMC|AEEC Welcome Speech Presented during AMC Opening Session ORLANDO

Monday, April 22, 2013 Marijan Jozic, AMC Chairman
Director, Avionics Shop Engineering
KLM Royal Dutch Airlines

My fellow engineers!

I am happy that we all survived the fiscal cliff. Glad to see you here in Orlando. On behalf of the AMC and AEEC steering committee and our host Gables, welcome to the 64th AMC|AEEC conference. Welcome to Orlando. I wanted to say also on behalf of Mickey Mouse, but our lawyers told me that Mickey Mouse is Walt Disney's intellectual property and they might ask a fee for using their trade mark name. I also cannot tell you to not go to Walt Disney World and stay at our great AMC|AEEC conference instead because that would make the lawyers warn me again. It would be a kind of establishment of cartel against Walt Disney World. So let's keep it simple and let's do what we do best and that is aviation stuff. I hope that you all had a great time yesterday at the Gables party. It was really great. I thank our Gables friends for giving us such a wonderful time. The most charming part of the AMC|AEEC conference is that it is a magnificent formula. 700 professionals came from all over the globe, meet for 4 days, and solve each other's problems. Many new ideas pop up, new friendships and a great network is created, and if you do it right, you go home with 100 or more business cards and many great ideas to save money for your company. Since the Anchorage conference, we have been working hard to keep the world fleet flying. In the first place, I would like to tell you that I am privileged to be here with you sharing the passion for aviation.

Honestly, when it comes to technique, aircraft performance, maintenance, and new developments, we share the same passion. There are no borders to keep us apart. We are cooperating and keeping the world fleet flying. We are engineers. The definition of what engineering is varies from country to country. In the US and Canada, engineering is defined as: A regulated profession whose practice and practitioners are licensed

42

and governed by law In some English speaking countries, engineering has been seen as a somewhat dry,uninteresting field in popular culture and has also been thought to be the domain of nerds. For example, the cartoon character Dilbert is an engineer. Several *Star Trek* characters are engineers. "Q" in the James Bond movies is an engineer. One difficulty in increasing public awareness of the profession is that average people do not have any personal dealings with engineers, even though they benefit from their work every day.

By contrast, it is common to visit a doctor at least once a year, the accountant at tax time, the pharmacist for drugs, and, occasionally, even a lawyer. It is not common to visit an engineer. In companies and other organizations, there is a tendency to undervalue people with advanced technological and scientific skills compared to celebrities, fashion designers, sportsmen, entertainers, and managers. Engineers develop new technological solutions. During the engineering design process, the responsibilities of the engineer may include: defining problems, conducting and narrowing research, analyzing criteria, finding and analyzing solutions, and making decisions. Much of an engineer's time is spent on researching, locating, applying, and transferring information. Aircraft systems are complex. LRUs are complex. Software is complex. The whole aircraft is extremely complex and we engineers manage to overcome all issues to make them fly.

They, airframers, and OEMs spend a few years to design and a few months to build the aircraft; we at the operator's side spend 20 or more years flying them and maintaining them.

There are challenges. Let's mention some of them:

1. Integration of Aircraft systems is the biggest step in a good direction, but also the biggest puzzle if you are not used to working with super integrated systems. Troubleshooting is impossible without at least a laptop. Modifications are impossible without aircraft manufacturers because only they control how the interfaces and software work together. The task of the system integrator is extremely difficult.

2. Avionics shops are in a big dilemma. The equipment installed in the aircraft is extremely reliable. It is flying forever. Each LRU has software

inside. To test the LRUs in the shop and to do level 3 repairs as we desire, shops must invest a lot of cash because to test computers, you need computers and the dedicated software. The dilemma is: high test equipment costs, low failure rate, and no flow of LRUs. Sometimes you have to invest half a million dollars for a test setup which is used just a few times per year. You need more than 100 aircraft in your fleet to be cost effective.

3. It was not a joke when I started the speech mentioning our friends' lawyers. It looks like they are taking control over our activities. Contracts are extremely important. Lawyers have managed to change our industry. OEMs are contracting their sub suppliers and airlines; airframers are contracting their vendors and sub suppliers; Flight simulator manufacturers are contracting airframers and OEMs; engine manufacturers are contracting OEMs, subsuppliers and airframers; and they all are contracting airlines and MROs. The ATE manufacturers are contracting OEMs, airframers, airlines, and MROs. It is important to have a contract but it is also very complex, especially if you are not a lawyer. Besides contracts, there are export licenses, non-disclosure agreements, end-user agreements, designee letters, delegation letters, authorization letters, and ITAR licenses. It looks like lawyers have started to control our lives. But here is the catch: they need engineers to establish the scope of every contract. They, lawyers, know perfectly how to define clauses about warrantees, late deliveries, payments, delivery time, exchange points, and penalties but they know very little about the scope of the contract. So here is the opening for us engineers.

Do you remember the story about going to the doctor, going to the pharmacist, going to the lawyer...? Well, here is an example of a new standard expression: going to the engineer. Engineers, more than ever, provide the data for each scope of contract. You as an engineer must be consulted otherwise the contract is not worth signing and the life cycle costs will explode.

4. For avionics shops, independent MROs, or airline owned MROs, it is a big challenge to stay in the business. It is a deadly competition at the aftermarket. Airlines should partly blame themselves. In the time frame of introduction of the MD-11 it was also a crisis. Airlines were desperately looking how to save some money. MD-11 is not to blame, but it is just a

44

reference that we can define the time frame (in the second part of the 80's). At that time the airlines proclaimed: "We will create variable costs from our fixed costs. We will outsource the maintenance of components and save some money. If we contract the whole package to OEMs, we can get a better price." That's how it started. Once you are there you cannot go back. My airline is the last operator transporting passengers in the MD-11. Even now, when we are phasing out our MD-11 fleet, our inventory department is pressing very hard to get some of the capabilities in-house for just one year because of high outsource costs. But there is no way back. That is exactly the reason those airlines that are traditionally doing repairs at the aftermarket are losing big pieces of pie each year. Can you imagine how difficult it is to step into component maintenance for extremely reliable and expensive components which can be tested only on expensive test sets?

5. The AMC|AEEC conference is surviving because of its unique formula. The threat for the AMC is company mergers, which creates mega carriers and mega MROs but automatically decreases funding for AMC|AEEC because of a fee cap. We are urged to find a new business model to keep this activity going. But remember one important thing: if we stop this activity, every one of us will be alone and there is no way back. As long as airlines make money, they will invest and everybody will be happy. Our passengers demand a low seat price and if we lose the market, we cannot invest in aircraft and components. Only by working together, airlines, OEMs and airframers at the AMC|AEEC can make it possible to keep the conference, exchange experience, and the world a better place. Hydro mechanical engineers are jealous because they don't have such great conference. Don't forget that we are unique.

6. Safety is the biggest challenge in aviation. We all strive to have the same amount of take offs and successful landings. Last year was the safest year in history. We all here in the room contributed to that figure. We all can be proud of making the numbers and managing and improving the target level of safety. That is not easy. Our fleets have flown more miles per year than ever before. Our aluminum and nowadays plastic birds (or should I say the composite birds) are having an extreme high utilization and are hanging somewhere in the air for many hours. Due to

our extremely high engineering skills, we manage to lower the amount of fatalities. That is probably the biggest challenge of all.

7. And last but not least: the challenge of lifting up the image of the engineer. It is good to know who we are. A Google hit says an Engineer is someone who solves a problem you didn't know you had, in a way you don't understand. Some also say about the engineer:

You're Really an Engineer If:

1. you take a cruise so you can go on a personal tour in the engine room.
2. you see a good design and still have to change it.
3. your wife hasn't the foggiest idea of what you do at work.
4. you've tried to repair a 5 dollar radio.

But who are we really?

No engineer can walk away from an unsolvable problem until it is solved. No illness or distraction is sufficient to get the engineer off the case. These types of challenges quickly become a personal battle between the engineer and the laws of nature. Engineers will go without food and hygiene for days to solve a problem. And when they succeed in solving the problem they will experience an ego rush that is better than sex—and I mean the kind of sex where other people are involved. OK, nobody is snoring. That is always a good sign. Let's hammer the start of the 64th AMC|AEEC conference.

Notes from the Chairman 2.0

The Chairman wants to tell you something about Orlando

The Orlando conference has passed. It will be written in history as a great and smooth 64th AMC|AEEC conference. It was great to see you all in Florida. Although there was not much time to see the famous Mouse, we had a good time doing business in the hotel. The spouses had fun at Disney and our fun was fixing problems in the Avionics industry. Just to go through the conference highlights for people who were not there, we had almost 700 people registered: a lot of airline people and a lot of MROs, OEMs, and Airframers. Our host Gables held a pool party to socialize Sunday night. When you talk, eat, and drink it looks informal but we were actually already doing business. Almost all conversations were related to airplanes and avionics systems. It started with the classic questions:

How are you?
When did you arrive?
How is life?
Before the questions moved to business:
How is your company doing?
How many questions did you submit? Etc.

We all stopped socializing early and went to bed because we all knew Monday would be a long day and there were 3 more days after that. On Monday, the first day of the 64th year of conferences, the big ballroom was filled and we started at 9 o'clock sharp. I was very pleased and excited when delivering my speech, because you never know how the audience will react. When I was writing, I called my speech the 10 minutes of terror. But after the first 10 seconds, they became the 10 most enjoyable minutes in my life. Once in a lifetime you get the opportunity to say something to 700 engineers. I know from experience that engineers are not the easiest audience to entertain. After opening and that famous bang with the gavel we welcomed the keynote speaker. Once again Gables provided a very pleasant surprise in the great keynote speech. It was educational and interactive, containing a video with the interview of

Mr. Clark, the son of Gables' founder. I discussed Gables in the last issue of Plane Talk: "Gables, It's a Tradition." It is a one of a kind company and if we are an airline, OEM, competitor, or partner we must cherish it. In the modern world with takeovers, name changes, and many other business transactions, Gables proudly showcases the real meaning of tradition. Herewith I would like to congratulate the Volare, Trumbull, and Roger S. Goldberg award winners. The awards are very well deserved and we all are

happy for our awarded engineers. Next year somebody from the Plane Talk readership will be awarded. We do not know who but we know for sure that there are many great engineers around who are giving their knowledge, talent, passion, and creativity to the aviation industry. News travels fast but for those who are not aware, there are again changes in the AMC steering group. After many years of AMC conferences as an airline representative, Mitch Klink (my vice chairman) joined the ranks of the OEMs. During this last AMC we were joking about Mitch going to the dark side. Of course this was a joke; we were all happy that Mitch stayed in the aviation industry. People are changing jobs all the time but it is always a pity when they leave aviation after so many productive and passionate years for the "dot com" industry. We will see Mitch again at our next AMC and are certain that he will remain a great guy and a great engineer. After the official opening on Monday morning we started the open forum. We had 267 questions and a lot of work to do. But the longest journey starts with the first step, and so we began the series of questions with question number 1.

It took us 3 1/2 days to handle all of 267 questions. 90% were closed with success and we isolated 31 questions as "AMC Success Stories." I must give a lot of credit to the OEMs. They worked hard before the conference to solve problems and make each one a success. They were very proud when the moderator banged the gavel and declared an "AMC Success Story." Every bang means that we have made the world a better place. It is not just a show at AMC: another avionics problem is solved, another component will perform better, another aircraft will fly with fewer complaints, and that all contributes to lifting the level of safety. Globally, it is just a tiny little part of the solution but if problems are not solved they will remain a safety risk. Tuesday evening was the AAI reception. In my eyes it is one of the highlights of the conference. Many small OEMs and MROs provided a great show. Outsiders cannot imagine how many terabytes of data is exchanged between engineers.

At the AAI reception an engineer goes where no man has gone before. He discovers new opportunities for business and new improvements. On Wednesday morning there was a one-hour airline only session. Well, we call it an airline only session but more a more appropriate term would be caucus. From Webster's Dictionary, a caucus is a meeting of supporters or members of a specific political party or movement. In our case we wanted to discuss a few things with airlines:

If airlines would like the idea of speed dating with OEMs. Computer based training of ARINC standards. The future of the AMC conference. Charging airlines for attending the conference. The airline only session will be discussed at the next steering committee meeting in October.

During this AMC conference we had 3 symposiums:
Monday: Engineering and Maintenance of Connected Aircraft
Tuesday: Data Harmonization
Wednesday: Solder Joints

The moderator asked attendees for a show of hands for those who do not have any No Fault Found (NFF) problems. Not one airline raised a hand. That was exactly the reason why the NFF group was given time to provide some information about their work. They are doing a great job and I am certain that we all will have a lot of advantages in our operation after they publish their work. It is always a great feeling that they fight so enthusiastically for a good cause. Thank you NFF group for doing a great job.

Last but not least I have to mention the great hospitality suites. There is no doubt that we all worked very hard to make the AMC|AEEC conference a success. We spent many hours formulating and collecting questions, and our friends from OEMs spent many hours answering them. Sometimes they had to fight for us in their own organization and eventually they fixed our problems. After so much hard work it is not a bad thing to have a celebration. This year was no exception. Esterline CMC Electronics has announced that Toronto is waiting for us. Next year we all will go to that great Canadian city to attend the 65th year of AMC|AEEC conferences. In cold Canada we, avionics engineers, will make Toronto the hottest place in the universe and our host will be waiting for us. I am proud to be an avionics engineer because an engineer is someone who solves a problem you didn't know you had in a way you don't understand.

The Best Way to Predict the Future is to Create It.

It is strange time in the aviation industry. Many things are happening around us and it is difficult to keep our focus on the ball. The fact is that we are in the middle of immense changes. Those of us who have attended recent AMC and AEEC conferences have felt that numerous changes are at work. A subtle, yet major, change has to do with our maintenance processes. Traditional aircraft will continue to employ traditional methods of maintenance. But new aircraft types are not only bringing new technologies but new interaction between airlines, MROs, and OEMs. The root cause of this shift is the extraordinary price of developing new aircraft systems. The B777 and A330 were last airframe types that used traditional LRUs in combination with software parts. It was and remains an example of great engineering design, as well as great equipment! Modern aircraft are increasingly software driven opposed to traditional LRU design. New airframes, namely the Boeing B787 and the Airbus A350, are firmly situated in quite a new world. Firstly, the scale of integration is huge. Only a few traditional black boxes are used. Engineers have decided to switch to card files, high speed data busses, fiber optics, and a lot of software. The fact that such complex machines are able to fly is a huge testament to the industry professionals I am proud to call my colleagues. Therefore, I can only say, "*Good job.*"

Never in the history of aviation (at least civil aviation) has somebody created more complex machines. Millions of engineering hours were spent to make it happen. Many OEMs who were previously competitors joined forces and made it possible. The successful result is the product: safe, efficient modern aircraft. In actuality, the design engineers created the future. And now the catch! Those millions of hours were not free. To be able to create the future, Boeing and Airbus had to mobilize great engineers who do not work for free. That is to say, their time and intellectual effort is not free. And we all understand that the millions of hours spent to design the aircraft costs a great deal of money. The business model used in any industry is for profit, as return on investment. This model obviously says that companies that have spent millions of man hours to design and build complex systems are looking to recoup their investment in time and resources. In a free market, the relationship of investment to return is somewhat linear. Simple systems usually result in proportionate simple return, and the time to earn that return is also proportionately short. But the case of expending resources to design and build complex systems changes the model. Complex systems require a substantial investment. The company is looking for a fair, but proportionate, return. But the difference is the time that it will take to earn the return on the investment. The industry is facing a growing shortage of engineers over the next 15 years. The new business model is such that the research, development, design, and build is not covered merely by the sale of the product. The logistics of the entire aftermarket must be used to get the return of investment. This is the new concept that is difficult to comprehend in our industry. The engineers and maintenance professionals that attended recent AMC conferences have seen these changes. They have gained an understanding of the concept of a longer time to recoup an investment, and how it involves the aftermarket of repair. But the average manager that does not attend AMC is totally surprised when he finds that some things are not working the traditional way anymore. Up until 2005, anyone was able to perform maintenance for anybody else. Third party work was simply considered normal operation. Shop resources were sufficient and there was always room for just one more customer. CMMs were readily available and you could order parts without problems. OEMs were more cooperative in this maintenance model. Getting a CMM was just a matter of sending an e-mail to your buddy at the OEM and he would reply to the e-mail with the CMM as an attachment. Of course, everybody still complained that the airlines were still not as profitable as they should be. But what's new about that? With the introduction of the B787 it has become a whole new ball game. And many people in the industry are still not aware of this new ball game. In the first place there is a new breed of people penetrating in the industry:

lawyers. Once they took over in some leadership roles at OEMs the airlines found themselves in trouble. In response, the airlines had to hire more lawyers too. That's how the ball game started. Just let me define lawyers based on the following story:

A physician, an engineer, and an attorney were discussing who among them belonged to the oldest of the three professions represented. The physician said, "Remember that, on the sixth day, God took a rib from Adam and fashioned Eve, making him the first surgeon. Therefore, medicine is the oldest profession." The engineer replied, "But, before that, God created the heavens and earth from chaos and confusion, and thus he was the first engineer. Therefore, engineering is an older profession than medicine." Then the lawyer spoke up: "Yes, but who do you think created all of the chaos and confusion?"

And here we are: engineers controlled by lawyers. My guess is that the whole new business model is not created by lawyers but by economists. Economists defined the top level philosophy and calculated the profitability. Then they decided to not to predict but to create the future by assigning lawyers to make sure that the model will be sustainable. So to keep model in shape, they created documents not familiar to engineers. This is the moment that the paper chaos started. Ten years ago we had Product Support Agreements (PSAs) but not a single engineer knew about it or saw it. Today it is the major topic of all discussions in aviation. But wait, there is more. Lawyers created the concepts of delegation letters, authorization letters, nondisclosure agreements, end users agreements, export licenses, tri-party agreements, intellectual property agreements, etc.

Of course the lawyers who are reading this will start to Laugh Out Loud (LOL) because it is not new. They have written these agreements for many years. I would not be surprised that even the ancient Romans had some kind of intellectual property agreement when they were building the aqueducts. But aerospace engineers are not used to working in these constraints. Not yet. Therefore, we all need to be aware of this new ball game, and at all levels. Average engineers need information to design the modification, test set, or software. But suddenly they are not receiving the data or information they need. Their buddies at the OEM are also not used to fighting the legal paraphernalia and they say (sometimes too easy): "Oh, it is intellectual property. You will not get it." They frustrate the airline engineers; there is tension! In the good old days we all were working together to keep the airplanes flying. Now we are annoying each other by sending formal documents to sign. Engineers send those documents to their lawyers and they say: "No, don't sign!" The engineer is frustrated because his projects are collapsing due to the growing paper machine. He knows that everyone will look at him expecting to finish the test set, software, or modification and they will not take no for the answer. It is easy to say: "If this is the future, I don't need this kind of future" and throw the towel in the ring (give up). While this is a solution (and engineers love solutions), it is definitely not the correct one. Engineering data is an often overlooked provision in many negotiations.

A better solution is available. It is time that engineers start to learn some things regarding the future. Therefore, the AMC created the SCEA (Standard for Cost Effective Acquisition) Working Group. They have produced the newest ARINC document, **ARINC Project Paper 674:** *Standard for Cost Effective Acquisition for Aircraft Life Cycle Support*. It is so new that it is not even adopted yet, but it will be in a matter of weeks. If you are an AMC Airline Member Organization, download it free of charge. If you are not, consider joining the AMC or at least purchase the standard. And read it. It will give you the answers. It really is that good. One of my airline's purchasing directors read it and afterwards she said, "This is exactly what we need. We will use it to train our buyers." The document tells you why you should do what you have to do. It is great material. Engineering people should do the same. Read it and be familiar with it just like you made yourself familiar with ARINC 429. Read it and believe me when I say, "It will make your life easier." That is the start to creating our future.

My Buddies and me in Zagreb

We just came home from the AMC|AEEC mid-term meeting in Zagreb. Zagreb is the Capital of Croatia and we decided to go there because we wanted to promote ARINC Industry Activities as well as attract local airlines to the AMC|AEEC Conferences. If you look at a map, you will see that Croatia is quite a big area and if you do some research, you will find that there are a number of airlines operating there. They are fairly small and you cannot compare them with the western world. It is obvious that they are not affected with mergers and acquisitions. However, that soon may be a reality. In local Croatian papers, you can read about how the Chinese are investigating the possibility of purchasing local airports or how Asian operators are negotiating a takeover of local airlines. Although many small operators are operating locally, they will soon face globalization. We sent invitations to 160 people, hoping that 10% would show up, socialize with us, and be present at the AMC|AEEC meeting. But just 7 were there, which is only 4.375%. That is a bad sign. This number has three explanations. In the first place, it could be the case that we sent invitations to the wrong people, which I doubt because every e-mail address was double checked on LinkedIn. The second explanation could be that they all are in crisis and therefore did not send a representative to the meeting, although we were in their region. The third explanation worries me because it means that they do I am sure that they know about ARINC 429 busses and about ARINC 600 connector specifications, but it is possible that they do not know about standardization activities and about the AMC|AEEC conferences. That is definitely not good. It means that we all should do something to make ARINC Industry Activities visible to the whole world. In Zagreb we had two separate meetings: the AMC and AEEC mid-term leadership meetings. Our AEEC friends were working very hard to make sure that a few new standards were adopted and that some older ones were updated. The AMC Steering Committee meeting started with a handicap. Two members were stuck in Paris and Hamburg and arrived a day late. That is the price you pay if you go to a relatively remote area. Croatia is at the edge of the European Union and there is only one flight each day to and from European destinations. If you miss the flight, you have to wait 24 hours. That was the reason that we were not able to vote for adoption of the Standard for Cost Effective Acquisition (SCEA) document on the first day of the meeting. On the second day, SCEA was unanimously adopted. We are proud to announce that ARINC 674 is now an official ARINC Standard.

Just a little bit about the AMC mid-term meeting. We reviewed many topics to figure out how to improve the conference. Of course we analyzed the surveys because the voice of the customer is important, and the customers are our conference attendees. The conference was rated as 8.6 on a scale between 1 and 10. Since we got 11% feedback, the mark 8.6 looks reliable. The mark is high, but there is still room for improvement. Generally speaking, we should be very satisfied but we are engineers and engineers strive to reach perfection. Therefore, we are aiming to get higher marks at the next conference We also spent some time selecting the topics for the

2014 AMC Symposiums. One of the items will be 3D printing of parts for aviation. It looks like we need to bring to attention to 3D printing. Some OEMs and even airframers are using 3D printers to make parts. We must explore 3D printing. Even the International Space Station (ISS) has a 3D printer to make parts if there is a desire. In publications, there is a lot of attention for 3D printing. We should not just sit and wait but be part of it. Therefore, if somebody has experience with 3D printing and wishes to share it with us at the upcoming AMC conference, let us know. Please contact the AMC Steering Committee or me directly at m.jozic@klm.com and I will provide the details. We need three speakers about 3D printing. Each should speak for 25 minutes and tell us about his or her experiences and discoveries. We are at the beginning of a new chapter in aviation history. Be a part of it and show us your pioneering work. Besides a lot of topics we handled at the mid-term meeting, there was time to socialize and learn about Croatia. Croatia is not a well-known country, but if you explore it you will learn the history in that part of the world. There are people known all over the world that nobody knows have Croatian origins. For example, there are many Croatian engineers who invented items we all use today:

For example, Nikola Tesla (you all know about A/C current) was born very near Zagreb. The inventor of the ball point and fountain pen, Slavoljub Penkala, placed the first ball point factory in Zagreb.

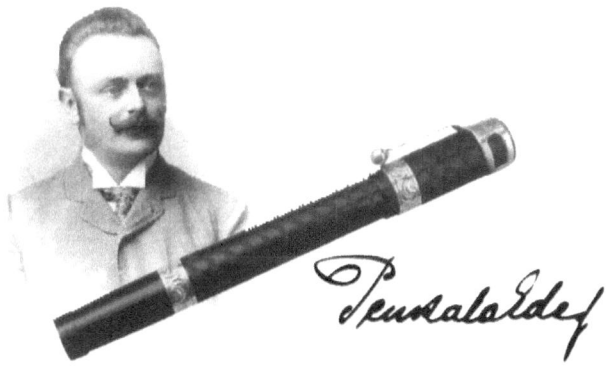

The inventor of MP3 code, Tomislav Uzalac, was from Zagreb and studied at the University of Zagreb. He was the creator of the first modern MP3 audio engines, a piece of software that translates a digital music file into analog sound. The first torpedo was invented by Ivan Lupis in the city of Rijeka in the 19th century. The famous Moholayer (or Mohodiscontinuity) of the Earth is well known among seismologists. Andrija Mohorovcic's discovery was essential for understanding the inner structure of the Earth and the behavior of seismic waves (Mohorovcic = MOHO).

Mohorovcic was from Zagreb. There is also Croatian Antony Maglica, inventor of the Mag-Lite flashlight, one of the top 100 products that "America makes best." David Schwarz created the first flyable rigid airship, which was also the first airship with an external hull made entirely of metal. He died before he could see it finally flown; he flew the prototype just one meter above ground. He died at the age of 42, just three days before maiden voyage. Some sources have claimed that Count Ferdinand Graf von Zeppelin purchased Schwarz's airship patent from his widow, a claim which has been disputed . There is also neck tie: the French word for tie is cravate, stemming from croate meaning Croatian. 17th century Croatian soldiers were the first to wear a tie to distinguish themselves from other armies. So you can see — Croatia's history has affected world history. And recently, Croatia affected the world's aviation industry by hosting the engineering professionals from the leadership of the AMC and AEEC. The conclusion about Zagreb mid-term meeting would be: Just another successful AMC | AEEC mid term meeting.

Jeff Kimball has Left the Building!

Just recently, we got a message that our well known buddy and AMC celebrity is retiring in December 2013. I have met many people in my life, but I consider Jeff one of the most remarkable. For many years, we shared the good and the bad times in aviation. Although we live on different continents and we work for totally different companies, there is a common interest which brings us together. The common interest is solving issues and making this aviation world a better place. Herewith I would like to thank Jeff for his effort and enthusiastic attitude. Jeff is a constellation of knowledge. The best part of it is that he was always willing to share that knowledge and to give you much more information than you asked from him. Only the greatest people in the industry have that ability. Only those who share knowledge can count on cooperation and mutual respect. This is what makes him remarkable. There is one more common thing which connects us. That is the ability to see the humor in engineering activities. Engineering stuff is dry and serious, data driven, mathematical, and statistical, but there is a lot of humor everywhere. Jeff is one of us who recognizes the humor and remembers it. That is why we all enjoy listening to the many stories that Jeff relates to us. Thank you for your stories, Jeff. They were always interesting and Jeff was always able to see and to emphasize the humor in even the sometimes serious stories. Approximately a year ago, we were in Amsterdam during a SCEA Working Group meeting. The meeting was on the day of St Nicolas. St Nicolas Day is a celebration in some European countries where St Nicolas (Jeff called him St. Nick) travels around and brings presents to people. It is like Santa Claus but on a different date: the 5th of December. Normally you do not see him when he enters your house through the chimney. He has to visit about 100 million houses in one night (12 hours). That means that he stays in each house 0.3 seconds. It is a blink of an eye. Therefore, you do not see him. But KLM arranged that St. Nick appear in the room where we had the SCEA meeting. Jeff was very surprised by St. Nicolas. We all remember that moment. St Nick asked Jeff a question about airframe manufacturers and their relationship with airlines and we all were a bit surprised that Jeff did not have an answer or story to tell, which was very unusual! At the 2013 AMC in Orlando, we surprised Jeff by playing a video about the world's greatest business mind. That was one of the topics of the first day of AMC in Orlando. Jeff was again flabbergasted and we were again surprised that there was no comment or long story to tell from his side. Jeff was able to accept jokes about him. It is an ability that only the greatest guys have. We all can learn from it. Thank you!

The 20th of December marks the first day of the rest of Jeff's life. He deserves retirement. I hope that Jeff will be able to slow down and do things he likes. One thing is sure: The AMC conference will not be the same without Kimball. Boeing will not be the same without Kimball. The aviation industry will not be the same without Kimball. But starting tomorrow, we will try to pick up the bits and pieces and continue. It was a great privilege to work with Jeff and to enjoy Jeff's presence at the AMC. No doubt we will meet again but until then, enjoy the life! ....AMC Chairman......

Sisyphus was an Engineer!

In Greek mythology, Sisyphus, the king of Ephyra, was sentenced to an eternity of frustration and wasted efforts because he believed his cleverness surpassed even the greatest of the gods. Every day, Sisyphus would roll a huge, enchanted boulder up a steep hill. Before he could reach the top of the hill, the massive stone would roll back down, forcing him to begin his efforts again. As he pushed the rock up the hill, it occurred to him that life was a journey, not a destination. This story is over 3,000 years old. Sometimes authors write a story about a subject but in fact they mean something totally different. There is symbolism in stories and symbols have different meanings for different people. I am certain that I have broken the secret code of the Sisyphus story. Picture this:

Once upon a time, there was an avionics engineer working in an avionics shop. The company just purchased a fleet of B787 aircraft. The avionics engineer received a list of part numbers from the airframer. The list was called an RSPL. He isolated a few part numbers and decided to start capability development.

His superiors asked for planning, and based on his experience with legacy aircraft types he said, "We should implement the 4 M's (Machines, Material, Methods, and Man). Let's be careful and add some time to the development of capabilities process. OK. It will take 6 months. Since I cannot work on my project 100% of the time, I can do 4 or 5 projects simultaneously. In 6 months the capabilities will be up and running." Just like he used to do before, he sent an e-mail to his buddy at an OEM asking for a CMM. The answer was: "I have no

61

authority to send you the CMM because it is our intellectual property. Our tech pub department will help you." He contacted them and received a short email with an attachment called an End Users License Agreement (EULA). If he signed it and sent it back, he would get the CMM. OK, that was a setback. He had to go back to his legal department because the EULA was full of legal stuff which was written in small letters. It smelled a bit strange. Legal checked it and said "No, we cannot sign this. We need to change the wording in some places." Then, he returned to the OEM and their legal department said "No, we do not agree. We want it different." One more setback; he goes back to his own legal department. Eventually, the document is signed. After some time the CMM was delivered. Our engineer immediately noticed that CMM was very thin. He could not do anything with it, but the CMM referenced a Technical Support and Data Package (TSDP). This was one more setback because the engineer had to go back to the OEM, back to the EULA, back to legal, etc. After 4 months he received the required data. One M is done. 3 M's to go. Of course, our engineer is not stupid. He was not sitting for 4 months; he was handling the other M's. Material: Material means initial provisioning. That is material required in the store when the shop is doing maintenance and repair. Our engineer is supposed to analyze the IPC of the unit and select the parts which should be purchased for the first shop visits. He can buy all the parts but there is a chance that some of the parts will never be used. If he reads the contract, he can find the clause which gives him rights to request the list from the OEMs with recommended piece parts. If he asks them, they will typically say "No, we do not have it." One more setback. It takes some time before they start to cooperate. Then he gets the list and a notification that their lead time is 36 months to deliver the parts. One more setback. After some time our engineer decides to purchase the unit from a second hand market and perform a teardown. He will obtain all the parts and it is much faster than 36 months. In theory, he can repair only one LRU because he has just one part of each. So he will still have to order some parts and wait 36 months. This is one more setback. Man: This used to be easy. You would just ask the OEM representative when the course was being held and you would send people to the course. This time it is different. There are no scheduled courses. Everything occurs upon request. And even worse, the courses are very short: 2 days, 3 days if you are lucky, but not more. Now we are dealing with extremely complicated avionics LRUs and the course is just 2 days. What can you learn in 2 days? A short introduction? Theory of operation? Quick peek in the books and schematics? Visit the repair facility and if the avionics engineer is lucky, he will actually see the LRU. This is a potential setback which will be initiated later. Once the unit is in his shop, he will have to learn hard way. Machine: This is actually the test set. If we are talking about a B787 test set, then we are facing one setback after another. In the first place, it takes 6 months or more to get a quote for the test set. Even if the quote is provided, it is not complete. Some items are not quoted, some items are missing, and some items are not required but quoted. So there are a few setbacks before the quote is complete. The next surprise is the extremely high price. The price is called quote to kill. If our engineer survives the attack of anger, he can calm down and decide to look in the books and try to figure out if he can design the test set in his own facility. Can you imagine? What a setback! But he is the avionics engineer and he is passionate. He can do it. After some time he will be able to see the light in the tunnel. Of course there will be a series of setbacks in the development process, but a year later the test set is operational. Material is in his store. The technicians are trained. The latest revision of CMM is on the shelf and he thinks he is done. The only item which is not there is the LRU. Then our engineer realizes that the LRU is outsourced under a long contract (typically 10 years nowadays) and that there is no way to get it in his shop. He was dealing with purchasing trying to get the parts, training test sets, etc.,ready while another guy in the inventory department was dealing with other purchasing people who were writing contracts for outsourcing. Setback? This is a major setback. But the engineer started to dig in the data and contracts and discovered that there was an opening in the contract saying that if the shop established the capability, it might be possible to deviate from the contract upon mutual agreement. The OEM is, of course, not willing to change the situation and they admit openly that they are going to lose revenue if they allow him to do the repair in his shop. After a few setbacks, a change is added to the contract, signed by the relevant people, and he is ready to go. After some time he may realize that the LRUs are not arriving to his shop. The computer system is not updated with the latest data and the LRU is still outsourced. After that setback, the LRU finally arrives to the shop for repair. Instead of 6 months, it is 2 years lead time for capability. His new name is SISYPHUS.

So let's count all the setbacks - approximately 13, not counting a few series of setbacks. The engineer is the guy rolling the stone uphill. When he thinks that he has made it to the top, the stone rolls down (sometimes all the way). In modern English, it is called a setback. Sisyphus was punished by doing hopeless labor. The engineer is also punished by doing hopeless labor, knowing that many of his actions were nonexistent 10 years ago. The avionics engineer is consigned until his retirement, which is bad enough!

The final thought:

In experiments that test how workers respond when the meaning of their task is diminished, the resulting demotivation is called the Sisyphean condition. These experiments found that people work harder when their work seems more meaningful, and many underestimate the relationship between meaning and motivation.

Special Tribute and Hats Off to Professor Marijan!

By: John Leslie, NAASCO

As a youngster I was lucky enough to get a job in a component repair shop back in 1962. At the time I had no idea that getting just a job would lead to the American Dream and a lifelong career in aviation. Just apply simple math and you will see I have been working in the aviation component repair business for 52 years. During my lifetime I worked for 3 aviation companies, and then in 1984 I started NAASCO with help from two of our kids in a lean-to on the side of an old hangar on Long Island, New York. In the beginning we focused on helicopter LRUs, then over the years we found bigger opportunities in the airline business. To this day helicopters are a very big part of our business, but the airline side offers an endless stream of new opportunities for our company. As a business owner I am passionate about NAASCO and the solutions we provide for expensive or problematic LRUs. Then there is Marijan Jozic, who reminds me of Albert Einstein. Once he gets going it is evident he is passionate about a lot of things, including the AMC, Airlines, and Avionics in general. It bothers the heck out of him that aviation mechanics and engineers do not get the recognition and respect they deserve. Whenever "Professor Marijan" takes the podium you quickly realize he is an educator, leader, and a visionary who takes charge and gets right to the point. When he enters a room you know he has arrived because not only does he know practically everyone, he emits an aura of confidence. If you are on the receiving side of his friendship his warm smile puts you at ease. When he speaks you unconsciously pay attention and realize he is not only passionate about Avionics and KLM, he is sincere about anything he talks about.

As an engineer, Marijan leads by example and advocates education to keep up with the technological advances that make flying safer. The manner in which he recognizes trendsetters and industry leaders is one of his best attributes. An example of his passion to recognize industry leaders is evident by his special whimsical Power Point Roast announcement and testimonial at last year's AMC by recognizing Jeff Kimball of Boeing as one of the smartest men in the world. Until Marijan elaborated just how many things Jeff has accomplished, we would have never known how brilliant he is. I am sorry to say I did not meet the Professor until 1999 at the AMC in Baltimore, Maryland, I approached him about our Sil-Met™ ATA 24 Power Relay and contactor repairs, and he immediately took notice and started asking questions about NAASCO's new niche market. A passion for the industry and for cost saving solutions absolutely drives Marijan. So he and a colleague flew to New York to see what we could do for KLM. He has been a good friend and customer ever since the visit.

In order to relieve Marijan's over-packed memory bank, I think this crazy Croatian and Guarding Angel of the AMC felt compelled to write a couple of books to relieve his brain pain and make room for the technical advances yet to come. I do not think you can hold all that information in your head without letting it out. I recommend everyone in the aircraft maintenance business read his two books: I, Avionics Engineer (2007) and You, Avionics Engineer (2012). Having put only some of his thoughts in writing must have truly cleared his head, and I bet his books could conceivably be used as textbooks for those entering the industry - the equivalent of taking a college course on aircraft maintenance. Aircraft maintenance instructors or other professors could certainly learn more about their profession and the evolution of avionics in modern airplanes. If they do, they will understand why I think he is an exceptional person and deserves to be recognized for his influence and view of the industry.

NAASCO NORTHEAST CORP.

Welcome to AMC. Welcome to Toronto!

On behalf of the AMC Steering Group, we look forward to seeing you for the 65th annual AMC | AEEC in Toronto, April 14-17, 2014, hosted by Esterline CMC Electronics.

The aviation industry has always had a rough time, and last year was no exception. New airlines were established, old airlines were changing names, and OEMs were buying OEMs or selling some of their subsidiaries. It is a lot of activity. The wind of change is always blowing and last year we all felt that breeze when we heard that ARINC was purchased by Rockwell Collins and just one hour later our very well-known ARINC Industry Activities (IA) was sold to the SAE ITC. SAE is a global association of more than 138,000 engineers and related technical experts in the aerospace, automotive, and commercial-vehicle industries. SAE International's core competencies are life-long learning and voluntary consensus standards development.

The great news for AMC is that it is business as usual! The biggest worry for Industry Activities was independence. If an OEM owned ARINC IA, independence would not be granted. Now under SAE it stays independent, which is good.

We sincerely appreciate the continued support from our Member Organizations and Corporate Sponsors of Industry Activities that enables us to conduct this one of a kind global event for the benefit of all our attendees. As an operator attending AMC, please spread the word about who we are and what we do. As a supplier or airframer, please remember that being an AAI member is an important contribution to AMC, but if your organization has experienced increasing success as a result of participating in AMC, please consider also becoming a Corporate Sponsor of Industry Activities. By growing our memberships and sponsorships, we can ensure the future of AMC as your primary source for constructively resolving technical issues, providing information exchange, and networking with business partners you might only see this time of year. The scope of our activity has not changed under new ownership.

Now more than ever, we must stay strong in our engineering activities and help each other provide safe and reliable air transport in a cost effective manner. The success of the OEMs depends on the success of the airlines – we are customers, not necessarily competitors. The only way for an OEM to make revenue is through operators. If we, operators, are not buying components and airplanes, we will not be successful. We cannot ignore the commercial aspects of our business knowing that each of the key players struggle for their slice of the pie. Only through cooperation at AMC can each of the players (OEMs, MROs, airlines, and airframers) help one other to succeed in the business and stay in business.

Saying that, I would like to remind you that Toronto is the place and April 14, 2014, is the time. It can be cold in Canada, but the Toronto-Sheraton will be the hottest place in universe accommodating the 65th AMC | AEEC conference. Be there or be square.

2014 AMC Toronto—Another $ucce$$ $tory!

For over 65 years, airline maintenance professionals and their counterparts at their supplier companies have been coming together annually to resolve the problems in commercial aircraft avionics that have proved difficult to solve at home. Keeping the electronics that assist in that, flying operating at the lowest possible life-cycle cost is a piece of the puzzle. Every dollar saved in maintenance is a sorely needed dollar on the bottom line of an airline. Every mechanical delay or cancellation avoided pays benefits. And so year after year, in good times and bad, the avionics maintenance industry comes together in the spring to help the bottom line. The times are tough and profits are low. If you look at the distribution of costs for average airline, it will give the following picture:

Fuel costs: 29%
Salaries: 20%
Ownership costs: 16%
Government taxes and fees: 14%
Maintenance costs: 11%
Other costs: 9%
Profit: 1%

This is not my own calculations to make it interesting, but a calculation published in The Wall Street Journal. I can tell you it is aggravating. Engineers always want to have a feeling for the numbers. If The Wall Street Journal is telling the truth (and we all know that they are), we can use those numbers and calculate how many seats we have to sell in the airplane to make that lousy 1% profit. Fuel costs can account for

nearly 30% of an airline's total costs. Here are the numbers:

What you can see is that if we sell all seats on our flights, only a few seats will make the difference. In other words: 4 seats in our B747-400s and 6 seats in our A380s are providing profit. B737NGs, A320s, and regionals must have all seats sold to be able to get profit from just one lousy seat. Isn't that terrible? And consumers expect us to lower the price of the tickets. How unfair is that? That is why I say it is exasperating. Well, our management knows what is making the difference. That is the line called Maintenance Costs. If you lower maintenance costs, almost instantly you can see that the profit will improve. One of the tools to lower those costs is AMC. Send engineers to AMC and you can gain results, lower the costs, and have a healthier airline. At AMC 2014, only 50 airlines sent representatives. Potentially, only 50 airlines can use AMC to lower the costs and make sure that profit will be generated with more than 1 seat.

	%	B747-400 350 seats	A380 580 seats	B737NG/A320 120 seats	Regional Jets 100 seats
Fuel Costs	29	102	168	35	29
Salaries	20	70	116	24	20
Ownership Costs	16	56	93	19	16
Government Taxes/Fees	14	49	81	17	14
Maintenance Costs	11	39	64	13	11
Other Costs	9	32	52	11	9
Profit	1	4	6	1	1

Many, although they clearly understand the value to the airlines, wonder what is in it for the well over 140 suppliers who make up three-fourths of the almost 700 registered attendees each year. The answer is that it pays for them as well. Many of the supplier attendees have repair shops. A solution helps them, too. The reason AMC continues to be so successful lies in its mode of working together to solve problems. AMC is not a place where the customer comes to "bash" the supplier. Also, it is not the place to discuss commercial issues and pricing! The questions are all available to the supplier in advance of the meeting and frequently the answers are brought to the meeting for the combined group to take away with them. It is this spirit of cooperation, not only between suppliers and their customers, but also between the companies whose marketing departments compete so fiercely for the business, that makes AMC unique. The AMC brings value to airlines and suppliers as well. It has already been 65 years, and the AMC formula is still working. Now and then we have to tune it a bit, but basically the conference setup is the same: to stay under the same roof for 4 days and solve problems. There are people who have attended AMC for many consecutive years. They will definitely not come back if there is nothing to gain. It can be just an answer on their problem that cannot be solved at home or a tip they hear in the corridor or during lunch. I can give you an example. In Paris (many years ago) I ran into a company during the AAI reception. My standard opening question was: "Please tell me something about your products." He said: "We are building test equipment and we are specialized in series of one!" He have made my day! I knew that we would be business partners. Today, we have a strong business relationship. It started in Paris more than 10 years ago and it is only getting stronger. Why? Because it is a win-win situation. We respect each other; we help each other, sometimes for free. And eventually, the piece of pie that we share gets bigger and bigger. That is the secret of AMC. To help the bottom line of their companies, the airlines send their maintenance experts to AMC. They pay the bill to get the savings. Maybe you should add your voice next year. Supporting AMC does not cost, it pays off. You can add one more seat that generates profit for your airline if you take a chance.

Make it happen. Make the difference. Make another success story.

Avionics Master Plan 2.0

Not very long ago (at the end of the 1990's), there was a dilemma about what to do with the world fleet. Many aluminum birds were happily flying around and engineers were having headaches. At that time, the majority of aircraft was flying without TAWS and without GPS. There were discussions about the necessity of GPS, compliance with RNP 5 or RNP 3 and, of course, about compliance with FANS 2 requirements. System engineers were under a lot of pressure because management wanted cost effective solutions. There were a lot of discussions about how to comply and what kind of equipment should be installed. GPS could be installed as standalone GPS or inside MMR. If you installed it inside MMR, you needed MMR instead of ILS, meaning that you eliminated three expensive ILSs and installed three even more expensive MMRs. A possible option was to buy an EGPWC computer with an internal GPS card. But then there was just one GPS on board and you needed two to comply with RNP 3 requirements. Only brave engineers were proposing a full blown and expensive system. The idea behind that is as follows: you invest a lot of money now, but your aircraft is capable of growing and accomplishing the next technological step (and the second step will be cheaper). This meant that on a B747-400, you should install the triple MMR (with GPS and ILS), upgrade FMCs and INUs, install EGPWS, and modify the EFIS control panel. Of course, you would need a lot of wiring changes and a lot of ground time to do it all. That project was called Avionics Master Plan (AMP 1). After AMP 1, you were able to add other things without breaking up or removing some parts. The aircraft was fully fit for next 20 years; even the next step was possible (FANS 2 if required), such as installation of Enhanced Surveillance and Extended Squitter (AMP 1.1), CPDLC (AMP 1.2) or EFB (AMP 1.3), and Cabin Surveillance (AMP 1.4). If somebody tried to save money on the AMP 1 project, he found himself in a situation where at a later date he had to spend a lot of effort and cash to catch up and be compliant with TAWS, ELS, ES, CPDLC, EFB, and CS. Now, 20 almost years later, we are at the edge of Avionics Master Plan 2.0. We are in a similar situation with connectivity, assuming that connectivity is the next step. There are so many options and so many wishes. The times are harder than 20 years ago and at this moment, almost every mod house and service provider has their own solutions. If an engineer goes to conferences or to OEMs and airframers, he will be even more confused because of the extremely high quantity of data and solutions available. It is tough to make a good decision knowing that the new setup or system is supposed to be used for many years. For example, EFB was introduced approximately 10 years ago on the B777 as an option. We defined classes 1, 2, and 3 and tried to do a solid engineering job. But now, 10 years later, pilots do not like those full blown solid solutions. They just use iPads. I have flown to Washington DC in the cockpit of A330. There was an EFB as a standard option in the table in front of the pilot. The EFB screen in the table was covered with a hard plastic cover and on the top of that was an iPad. Now we are in the absurd situation of having aircraft with $100,000 USD worth of EFB systems but pilots are not using it because the iPad is considered more intuitive. AMP 2.0 is not defined and we see the dilemma about connectivity. Cockpit connectivity and cabin connectivity! That is an even bigger dilemma, but one thing is for sure: nobody cares any more for voice communication. When we were introducing Satcom, everybody was expecting voice communication to be a big business. Inmarsat was expecting that new cool telephoning technology from the aircraft would generate revenue. They did not want to lower the costs to attract people and the result was that nobody used the expensive voice mode of Satcom. Some business cases were predicting 45 or 60 minutes of voice communication per tail number per day. After introduction, it was a lousy 5 minutes per tail number per day. Can you imagine: Spending $400,000 USD per airplane and having a system which is not used 99.5% of the time available. Paying for maintenance of the system and gaining $55 USD per day. And, as the final joke, $55 USD is supposed to be shared with Inmarsat, so the airline will receive $20 USD revenue per day. Try to calculate the payback. I do not dare to do it. How about the credibility? As the next step, Connection by Boeing has designed a superb system. High speed internet for hundreds of passengers! It was free for some time and everybody was using it. And when it came to paying for it, nobody was home. So if we like it or not, our passengers will use data (e-mail, internet) only if it is for free or for an extremely low price. Every other option is going to fail just like it happened with Satcom. Of course, many people know about these past experiences. Therefore, not many of them want to be the trendsetter and start the new adventure with connectivity. Of course! But maybe there will be an urgent need for connectivity. When we were working on Avionics Master Plan 1.0, FAA and EASA helped a bit by mandating TAWS. That was exactly the drive to act fast because there was a deadline.

Now at the beginning of Avionics Master Plan 2.0, there is no solid deadline. It is not safety that will dictate the development but economics. If there is no money for connectivity it will not happen unless it is obvious that the business case is a no brainer and that every invested penny will be returned within a reasonable time. The bottom line is that Avionics Master Plan 2.0 will happen sooner or later. At this moment, it is difficult to say what the best solution is. There are too many variables and everybody knows that the first guys will have the advantage, but only for a short time. The new technology will come soon and it will be even better. There is no good direction and good solution at this moment, but somebody has to be first and start to lead. The following is still under construction:

Business model for the future: Based on the Satcom experience in Avionics Master Plan 1, it is not that easy to get a good model because due to Satcom, a lot of credibility was lost. Spending millions for a system which is not used will not happen again. Some airlines are claiming that they have connected aircraft, but we all know that something certified and flying by others does not necessarily mean that your needs will be fulfilled. Of course, you do not need something totally different, fancier, and cool because it can drive the costs very high – but do not over specify it. There is a difference between cockpit and cabin applications but perhaps introducing both can make your business case stronger and you might get both. Trying to get only cockpit or only cabin might be disaster and you will get nothing or delay it for a few years. The desire of your airline, routes your airline is flying, and passenger structure can perhaps make it possible to decide what system to use and what satellite bandwidth (service provider) can be used and be cost effective. What is good timing? Nobody knows, but one thing is sure: it is not too late. If there is doubt, the best possible way to eliminate that doubt is to visit conferences and establish your own plan of approach. The plan could be ambitious and it should cover the needs of your airline.

Once you know for sure what your desire is, it will be easier to define the concept and be aware that the airplane is not all that you need to modify. Even when airframers will tell you the airplane is ready, that is only the half of the story. Ground application is the second half and it can make your life even more miserable. That is the challenge! I bet ground application and software is where the money is. But that is another horror story. For Avionics Master Plan 2.0, the airplane installation is a piece of cake. Believe me.

Twenty Feet to Impact:
Without Test There is No Success At All

Testing in avionics environment today is a totally different cup of tea. We used to strictly follow the manuals. The manuals had a chapter which completely described the test setup. When you connected it that way and followed the test procedure step-by-step, you would have had success. With the introduction of the A320, B737NG, and then the B777 and A330, it was a bit different. You could buy a dedicated or universal test set, interface, and software and you could start repairing boxes and if lucky, earn a little extra money on the side. But then they launched the B787 (and the A350 is coming)! With the B787, a new business model was created and nothing is easy any more. In the new business model, OEMs are participating in the design of the aircraft. As a reward (actually a return on their investment), Boeing opened the doors for OEMs to penetrate into the aftermarket. The plan was established 10 years ago, but many airlines were sleeping. Traditionally, airlines and airline shops were doing equipment testing, repair, and overhaul. It is called aircraft or component maintenance.

The next step was to design aircraft equipment as a system and traditional LRU's (black boxes) were replaced by sensors, data concentrators, network servers, optical interfaces and many other items. New technologies were introduced (lead free soldering, Ball Grid Arrays (BGA), optical cables…the list goes on. Avionics shops were now entering a new world – a world of complicated and difficult technologies!

Disclosed below are some challenges with equipment testing in today's avionics shop:

Normally speaking, an avionics shop will use the Component Maintenance Manual (CMM) to build up the test setup. The CMM will also include the test procedure. Well, the new CMM's have no test setup information but have a reference to another document called the Technical Support and Data Package (TSDP). You have to order that document before you can figure out what to do next. The TSDP will also include the test procedure. Well, your first surprise is that you will need a Non-Disclosure Agreement (NDA) before you can get the TSDP. Your legal department will argue because most NDA's include the clause that you are not allowed to do third party work. If you disagree, the NDA will not be signed and there will be no TDSP and consequently, there will be no testing. Therefore, it is in everyone's interest to sign the NDA! The consequence is that your repair and overhaul business will narrow to your own fleet or you will pay a license fee for each LRU you tested for your third party customers. Ok, now you have spent 6 months of talking and signing the stuff and you can look into the test procedure in the TSDP as well as the test setup. First, what you notice is that you need something additional. You need the test software. The test software is a small piece of software that you load in the avionics unit you want to test according to the procedure specified in the TSDP. Without the test software, there is no testing. You use this software just for test. After test, the test software must be removed from the unit and the operational software must be reloaded in the Line Replaceable Unit (LRU). This story is repeated all over again with each new LRU: new NDA, ordering software, etc…

And yes, I almost forgot: there is a price for that test software part. Sometimes it can be as high as six figures in US dollars, and that is what we call a quote to kill. By now you have spent 1 year to figure out that this whole process is not cost effective. Unfortunately some OEMs have specified extremely long and complicated tests. That is because they simply took over the full blown qualification test they used to demonstrate to FAA or to EASA that the unit is airworthy. I can imagine that the test is required for certification, but every time the unit enters the shop? Why should you do such extreme and long testing? That is ridiculous. For example, picture an ordinary display unit on an ordinary aircraft must be tested and calibrated every time, even if the LCD and backlight is not replaced. The test requires you to measure colors from different angles, although every engineer knows that when you replace the LCD, you cannot install it under a wrong angle. But you need the machinery, which costs a half-million USD, to be able to position the display under a few different angles. This sensitive machinery is testing and adjusting colors for pilots' eyes, and is more adaptive to the colors than any machinery on earth. Sometimes the displays fly for ages, displays which will never pass the calibration test. The machinery and specs are so tight that almost every unit will fail and the consequence will be that a very expensive backlight or glass assembly will be replaced. In the good old days, we were able to fire up the LRU at the desk, do the troubleshooting, and then execute the return to service certification test. You cannot do this anymore! You need a full blown ATE with expensive ATE test time to do the troubleshooting. The problem with troubleshooting is that the trouble shoots back. It is difficult to pinpoint the component on the circuit board and you must repair by replacing the circuit board and outsourcing the repair to the OEM, which is extremely expensive. But they have test equipment such as flying or thermal probes and they can pinpoint the faulty component. So instead of replacing a resistor or capacitor like the OEMs do, the avionics shop is forced to pay a lot of cash for outsourcing the repair of the whole circuit board. Even if your avionics shop is able to troubleshoot all the way to a single component, many times those components cannot be purchased. OEMs will just say: Oh, the LRU is not designed for level 3 repairs. Period, end of story! Not so long ago, the OEMs were proud of building LRUs with extremely good Built In Test Equipment (BITE) functionality. It was a normal sales story to hear about BITE that covers 90% of failures in the unit. For example, the small circuit in the LRU will generate a calibrated test signal. The signal is injected at the input of the unit directly after the first filter. If the unit is working properly, the signal at the output will be of a certain value which will be analyzed and a green LED pass light will switch on. The avionics engineers are now happy. The idea was that you could power up the unit at the desk in the shop. Connect to power and some discretes, push the self-test button, and the unit will then self-test itself and the deal was done with a 90% certainty that if the unit passes the self-test, everything is good. Then you can connect it to an expensive ATE machine to do the full blown test. If it fails, you can troubleshoot the unit and perform repairs until it passes the self test. Can you imagine the savings of ATE time?

Well, many B787 components are compact and there is no self-test button on the component. The components are part of the system and a whole system test will show which component is bad. Now you bring the component to your shop and your test set must simulate the whole system. It is no wonder that test software and ATE Test Program Set (TPS) is so complicated and expensive. You can also choose to build the whole system.

Just buy 5 or 6 extremely expensive components and build an entire B787 system. Keeping in mind that you are on your own! You must design the hardware setup, design the test software, qualify the test, and keep everything up to date. That is even more expensive than ATE. Also we can argue how many times you do the ATE test. Let us say you run the test and it fails! Then you run it again 5 times and it passes every time. Can you release the unit as serviceable or should you run the test another 10 times. If it passes the test 10 more times, is the unit OK? If it fails 3 out of 10 times, what then? In the beginning I mentioned the new technologies which are normal for other industries but they have only recently started in the avionics environment with the B787. One of those technologies is the Ball Grid Array (BGA). A BGA is a type of surface-mount packaging (a chip carrier) used for integrated circuits. BGA packages are used to permanently mount devices such as microprocessors. A BGA can provide more interconnection pins than can be put on a dual in-line or flat package. The whole bottom surface of the device can be used instead of just the perimeter. After you attach it to the board and it is soldered, you must test the quality of the BGA assembly. That can only be done with an x-ray machine. So every decent avionics shop needs at least one of those machines. Besides an x-ray machine, new equipment needed in the shop is soldering equipment, microscopes (optical and with camera), test sets for optical cables, and many more things. The cost of test equipment and all software needed can be prohibitive for an airline to purchase. It is funny but very true that avionics shops experience a technology boost when a new aircraft enters the market. But due to many factors, avionics shops do not like to outsource old stuff. If you visit old avionics shops, you will notice two extremes. And old department with old fashioned stuff, with many test sets and breakout boxes, test cables, and seemingly museum artifacts. People working there protest vehemently if you want to scrap the old test sets and old stuff. I have seen a DC10 autopilot test set (Bendix test set) in a shop, 20 years after the last DC10 left the fleet. It was not used but the guys were guarding it passionately. They knew that it was worth a quarter million USD back in the 1970's but they refused to hear that it was worth nothing now. What can be said as a final thought? Testing in today's modern avionics environment is expensive. Test sets are expensive and components to be tested are extremely reliable. It is ingenuous to buy an expensive test set and use it twice a year. For decent utilization you need a decent flow of components. To create flow, you need a big quantity of aircraft under contract. If you decide to do third party work to utilize the test set, you must pay the license fee to OEMs. That means that your Return On Investment (ROI) will be longer and you will ask yourself: Am I on the best course?

It looks as if the glory days of avionics shops is over. There is no fun in testing any more. Avionics shops and LRU testing is moving to the OEMs whether we like it or not. Only simple capabilities with questionable value will be tested in avionics shops unless the shop discovers a big source of cash (which I doubt). It is difficult to accept that nice, complicated computers will not be done at airline avionics shops much longer. It is like jumping from the skyscraper and 20 feet before impact you say: "Hey, I am still alive, nothing is wrong at this moment." That is exactly what is happening with avionics shops. Most of shops only have 20 more feet, that is to say just a handful of standalone, independent avionics shops are little bit more that 20 feet from ground impact. We shall see!

Final Preparations for the 2015 AMC

As we all know, AMC will take place in Prague, Czech Republic. Prague is one of the most beautiful European cities. It once was part of the Habsburg Empire. It was in the Austria–Hungarian monarchy. These days Prague is capital of the Czech Republic. Well, in a couple of months Prague will be the hottest place in universe, because the AMC conference will be there. It is still a few months away, but it is starting to get serious. The AMC Steering Group will meet in Los Angles at the end of January to review all the open issues and align everything. Preparing a conference for 700 people is quite a challenge. The AMC Steering Group must take ensure that planning will be down to the smallest detail. Roles and responsibilities will be defined and divided between the steering group members. The details about the symposiums must be agreed upon, moderators must be selected, the hotel review will be completed, and speeches and timing should be finalized. All discussion Items for AMC will be reviewed and all unexpected planning items must be agreed upon and fixed. The interaction between the AMC and the AEEC must be determined and communicated between the leadership groups. I am always talking about the AMC, but our friends next door from the AEEC will have their annual General Session. We should not forget that they are doing very important work for the future of aviation. If they do their work right, our lives will be much easier. Therefore, never underestimate their activities. Saying that, I would like to change the subject and say something more about AMC and our core business, which is solving problems. We all know that the majority of our time during AMC conference we spend in the open forum session. That is the reason that we are there. We are the airlines! If airlines are there, the OEMs will be there. We need each other. OEMs want to have equipment that performs superb on our aircraft. We the airlines are the only party who can judge if equipment is performing as expected. If we provide the data to OEMs, they can and will improve the equipment.

Well, AMC is here to be the medium bridging OEMs, MROs, and airline shops and making this world a better place. The AMC is actually here to save the world. It all starts with formulating discussion items and questions for the AMC. The instructions are on the AMC webpage. Please read it carefully. Nevertheless, I want to give you some advice. In the first place, choose the chapter to place the question. If the question is very general, place it in the chapter called Avionics Philosophy. Otherwise, choose it by the ATA number. The next important subject is the Line Replaceable Unit (LRU) name. Do not forget to fill that blank. If the question is general, provide two or three keywords which describe the subject. The next blank is for an LRU Part Number. This is very important. If you leave it open, you may not get a satisfactory answer. The same is with the field titled Vendor. If you leave it open, nobody will work on your question and you do not want that. Put at least the name of airframer (Boeing, Airbus, Fokker, Embraer, Bombardier, etc.). If you leave it open, do not be disappointed if there is no response at the open forum. The next field is Aircraft Type. If it is more than one, put them all. Next, the ATA number will give the indication where to locate the question. The field "From" expects you to put your company name (3 letter code). The last field says: If MRO, the Associated Airline. Meaning, if you are submitting a question on behalf of MRO, you should put an indication of the airline(s) involved. The big open format is for your question. Be careful that your question is not just a statement, but that your question indicates what you actually want.

Some examples:

Airframer, supplier, and operator comments, please! Why has the OEM not issued a SB or SIL on this issue?

If this ADC upgrade is mandated, will it be a field modification or require a shop visit?

Have any other airlines determined this and come up with a solution?

Please provide us with the method of restraining lead repair!

Can the airframer please provide the appropriate approval
and a timeline for acceptance?

The quality of the answer depends on your question, so pay special attention in formulating your question(s). Spend some time because for every dollar you spend on formulating a question (time is money), you will most likely get back at least two dollars, meaning that it is better to spend some time creating a discussion item than to put your dollars in the bank. With present interest rates, your bank will net you 1.002 dollars at the end of the year. When you submit a question for AMC, you will get a double return of investment in just a few months. Think about that! Ok, we know now how to fill out the form and submit the question. Next step is to register for the conference and to obtain a room at the Hilton Prague. Act quickly because we expect 700 or more attendees. Make sure that you are in the same hotel because there is a lot of evening activities. There will be an AAI reception on Tuesday and a lot of hospitality suites. At this past year's AMC, I counted 33 hospitality suites last year. That means a lot of work in the evening. Make a plan and visit them all. You can meet with representatives and exchange ideas and business cards. Sometimes one good contact is enough to save thousands of dollars for your company. Do not underestimate these opportunities. It is your conference, it is what you make of it. Some large airlines are attending the AMC not because they are big, but because they see a lot of benefits and they want to be in the first line to get them. If you are an engineer of small airline, you can also reap these benefits and that can be huge for your company. I certainly hope to see you in Prague and also hope that you can submit a lot of good discussion Items. These items will not only benefit you but all of us −including OEMs. OEMs will help to solve the items to your satisfaction. Their interest is also to improve their product and save the world.

FINDING THE MONEY IN CONNECTIVITY

By Marijan Jozic *Chairman of the Avionics Maintenance Conference for ARINC Industry Activity and Capability Development Manager at a major global airline.*
Article for Avionics Magazine

Not very long ago, toward the end of the 90's, engineers worldwide were asking, "What should be done about the world fleet?"

Our aluminum birds were happily flying around without TAWS or GPS and it was giving engineers major headaches. There was controversy even about the necessity of GPS and compliance with RMP 5 or RMP 3 — and of course about compliance with the Fans 2 requirements. System engineers were under enormous pressure because management wanted cost effective solutions, but none were obvious. Discussions on how to comply and what kind of equipment should be installed proliferated.

Only brave engineers were proposing full-blown, expensive new systems. The idea behind it was that if you invest a lot of money up front, your aircraft upgrades would mean the next round of upgrades would be a lot cheaper and easier to execute. For example, on the 747-400 you would have installed the triple MMR (with GPS and ILS), upgraded FMC and INU's and installed EGPWS and modified the EFIS control panel. Of course, you need a lot of wiring changes and a lot of ground time to do it all, but such projects, the "Avionics Master Plan" (AMP 1), meant airline engineers were able to upgrade without totally deconstructing or removing systems at later upgrade dates.

AMP 1 also meant the aircraft was fully fit for the next 20 years, and that even the next step was possible, including Fans 2 if required, Enhanced Surveillance and Extended Squitter (AMP 1.1), CPDLC (AMP 1.2) or EFB (AMP 1.3) and Cabin Surveillance (AMP 1.4) without invasive maintenance work. If airlines tried to skimp out and cut upgrade and cost corners on AMP 1, the consequences were tangible in the long term. The money supposedly saved came back to bite as nearing deadlines meant intensive efforts and funds were needed to complete the massive suites of upgrades mandated, including TAWS, ELS, ES, CPDLC, EFB and CS.

Now, almost 20 years later, airlines find themselves at the brink of Avionics Master Plan 2.0 (AMP 2). This time the major prize is connectivity, and it will come at a great cost. There are so many applications, and so many wishes, and the economic times are harder even than 20 years ago. In the present moment, almost every service provider has a different solution. If an engineer goes to conferences or to OEMs and airframers, he will be even more confused because of the extremely high quantities of data and solutions available. It is tough to make a good decision knowing that the new setup or system is intended for many years of use.

The EFB was actually introduced approximately 10 years ago on the 777 as an option. We defined classes 1, 2 and 3 and tried to do a solid engineering job. But now, 10 years later, pilots don't like those full-blown solid solutions. They just use iPads. I have been flying to Washington, D.C. in the cockpit of an A330, and there is an EFB as a standard option in the table in front of the pilot. The EFB screen in the table is covered with a hard plastic cover and on the top of it an iPad and we are now in the absurd situation of having aircraft with 100k USD worth of EFB systems pilots are not using because iPad is cooler.

This sort of scenario means skepticism when it comes to the cool new stuff of connectivity. There is cockpit connectivity and cabin connectivity too, and to increase our cynicism we have in our recent memory the tragedy of voice communication. When we were introducing SATCOM, everybody was expecting voice communication to be big business. Inmarsat said that cool new telephoning service from the aircraft would generate revenue, but they didn't want to lower the costs. The result is that nobody is using the expensive SATCOM voice mode, and yet another expensive application is discarded, after some business cases predicted 45 or 60 minutes of voice communication would occur per tail number per day. The real number? Five minutes per tail number per day.

In the present scenario, Boeing has designed a connection system featuring high speed Internet for hundreds of passengers. It was free for some time and everybody was using it. And when it came to pay for it, nobody was at home. So, whether we like it or not, our passengers will use data (e-mail, Internet) only if it is free or comes at an extremely low price. Every other option is going to fail, just like with SATCOM.

Because of our missteps with SATCOM and EFBs, not many airlines want to be the trendsetter and start the new adventure with connectivity. But there is always the possibility of a mandate to save the day: when we were working on Avionics Master Plan 1.0, the FAA and EASA helped a bit by mandating TAWS. That was exactly the driver needed to inspire fast action, because there was a deadline. Now, at the beginning of Avionics Master Plan 2.0, there is no solid deadline. And it is not safety that will dictate development, but economics. If there is no money for connectivity, it will not happen unless it is obvious that the business case will be a no brainer and that every invested penny will return within a reasonable time.

So what is a good business model for the future? Spending millions for a system that is not used will not happen again. Some airlines are claiming that they have connected aircraft, but we all know that something certified and flown by others does not necessarily mean that mandate specs would be met. Additionally, choosing between cockpit and cabin applications is hard, but perhaps introducing both can make the business case stronger as trying to get only cockpit or only cabin connectivity might be a disaster, delaying ROI for years.

Nobody knows what "good timing" is, but one thing is sure and that is that it is not too late. If there is doubt, the best possible way is to visit conferences, do the hard thinking, and establish your own plan of approach.

The plan could be ambitious and it should cover the needs of your airline. Once you know for sure what the desire is, it would be easier to define the concept and be aware that the airplanes themselves are not all you will need to modify. Even airframers will tell you "the airplane is ready!" But no, that is only the half of the story. Ground application is the second half, and it can make your life even more miserable. That is the challenge! I predict ground application and software is where the money is. But that is another horror story, and the worst investment ever is not investing at all.

If there is doubt, the best possible way is to visit conferences, do the hard thinking, and establish your own plan of approach.

Matrix Reloaded

Written By: Marijan Jozic and Kevin Kramer

Often, when you discuss the subject of the AMC's benefits over and over again, your organization is still not able to express it in readable data. In the past, there were attempts to numerically express the benefits of the AMC conference. We asked attendees to provide the USD value for each question. Those USD values were added, multiplied by 10%, and then multiplied by the number of airlines and quantity of their aircraft. After some adjustments, the result would show the benefit of AMC, which was close to a hundred million USD. The results obtained that way were not considered reasonable. Therefore, discussions continued on the AMC Steering Group on how to provide a more sound approach to realizing the benefits of the AMC. If you try to improve the calculations, you will soon realize that it is extremely difficult to get numbers which are reliable, acceptable, and simple enough to be trusted. You must realize that the individuals who are financing your trip to AMC would exercise every number to challenge you to show what the benefit is for your company. They actually want to see the hard cash savings. It might be easier to say that you attend the AMC to prevent spending too much money for your company. Our financial people do not like the idea of preventative costs. Preventative costs are intangible for them. If you say, I went to the AMC, acquired an idea to change a procedure, and prevented spending $50K USD to purchase new parts by repairing something instead (thanks to a tip by an engineer in the lobby of the AMC hotel), they would not see the benefit of the AMC because you did not spend money in the first place. You can show that you are actually spending millions on the maintenance of an LRU and then went to the AMC to find that there is a more cost efficient Service Bulletin (SB) to address your problem. Then, the next year you can show the cost of maintaining your LRU dropped by 70% and the airline is not spending $200,000 USD per year but $60,000 USD. They will be satisfied. Next year, the finance people will of course ask for another 70% of cost reduction. But that is a different story.

Well, after years of those discussions, the AMC Steering Group has come up with a matrix that shows where to look for savings or spending prevention. The matrix is expandable in all directions based on your own experience and imagination, but basically, this is it:

INPUT	OUTCOME				
Savings at AMC based on:	SB/SIL	Contracts	Maintenance Concept Change	Operational Impacts (Delays, Cancellations)	Flight Safety
Operator Collaboration	X	X	X	X	X
Other Operators Experiences	X		X	X	X
Submitting Discussion Items	X	X	X	X	X
New repairs discovered		X	X		
Industry Wide Problems		X		X	X
New parts source tipped		X			
New services discovered		X			X
Historical AMC Discussion Items	X	X	X	X	X

The AMC provides the opportunity to enable positive outcomes.

The left column provides the basis of savings. It can be as simple as others' experiences or the outcome of your discussion items. You may attend AMC with just one target: to see if there is a new repair available for something flying on your aircraft. Also, you could come to the realization that you are not the only one with the problem and many operators are complaining about the same issue. You might accidentally run into another operator's engineer who shares with you a new source of parts which were declared obsolete. Other engineers might tell you about a shop (MRO) who is able to repair components which are normally not repairable. Do not underestimate the possibility that you might find the solution to your problem in one of the old AMC reports. The left column is clear. It is not limited to those 8 listed items. Use your experience and imagination to define more items. It would be interesting to know what new items you might define. There are 5 columns with the outcome of your efforts to save cash or prevent costs. Those columns are SB/SIL, Contract, Maintenance Concept Change, and Flight Safety.

Some AMC discussion items can have the outcome that a new SB or SIL is issued. Besides the Success Story, this is the item that can be exactly expressed in cash. If you accomplish an SB, it will cost some money but you will earn much more. This is an example of the saying: You must first spend money to earn money. It is up to your sound engineering judgment to figure out if you are going to do that and if the promised SB is actually beneficial. The column labeled Contract means that the outcome of collaboration at AMC could lead you to sign a contract for a service or hardware which should create a financial benefit for your company. A contract can be the result of different activities and again, it is up to your sound engineering judgment to figure out if you are actually going to do that or not.

The next topic is called Maintenance Concept Change. This topic might provide huge benefits just by changing the procedure or interval of maintenance. It could be that your people are doing something so stupid that nobody even dreamed about. You might discover it during a conversation at AMC. Again, low impact within your organization, but gigantic savings! That is exactly what you are looking for. That is exactly why you went to the AMC. The last column is called Flight Safety. You should ask yourself: does this issue affect the folks up in the cockpit? Could this component issue create a safety issue while in flight? Your Flight Ops department may be able to share with you some additional cost savings that you never thought of before (e.g., air turn-backs) that you can add to your AMC justification.

It is very possible that you can add columns with some additional outcomes or that you can express in numbers for your own savings. The matrix is designed to help you to justify the benefits of AMC. Do not hesitate to share your data. The whole purpose of AMC is to educate and share data to make commercial aviation safer, more efficient, and cost effective. An airline recently shared an example of how attending the AMC benefitted their company. At the 2014 AMC, another operator brought up a Weather Radar issue they were experiencing in their fleet. After the AMC, this airline reviewed their own reliability data and determined they too were affected by this issue. The related costs involved with this issue started to increase when they started looking at component repair costs and operational delays. The engineer brought this issue to his or her Flight Ops department and expressed concern about how this affected the Flight Crews and their use of the Weather Radar system in flight. None of this would have been discovered if the airline did not attend AMC.

Please feel free to share any justification ideas with the AMC Steering Group so we can build a comprehensive matrix that will benefit all our members. We hope that these ideas will assist you in justifying the costs of your trip to AMC and show your superiors and fellow engineers that AMC is the place to go if you want to make a big impact in terms of Avionics for your company.

m-by-n matrix

$$\begin{matrix}
a_{1,1} & a_{1,2} & a_{1,3} & \cdots \\
a_{2,1} & a_{2,2} & a_{2,3} & \cdots \\
a_{3,1} & a_{3,2} & a_{3,3} & \cdots \\
\vdots & \vdots & \vdots & \ddots
\end{matrix}$$

$a_{i,j}$ n columns j changes

m rows i changes

AEEC | AMC Opening Remarks Prague 2015

Good morning my fellow engineers! I hope that you all found some time to see the beautiful city of Prague. It is one of the most beautiful cities on the old continent. It used to be one of 4 major cities in Habsburg monarchy, together with Vienna, Budapest, and Zagreb. Each of those cities is now in a different country.

On behalf of the AMC Steering Group, welcome to the 66th AMC | AEEC conference! Welcome to Prague! The most charming part of the AMC | AEEC conference is that it is a magnificent formula. Just to remind you: 600 professionals come from all over the globe, meet for 4 days, and solve each other's problems. Many new ideas pop up, new friendships, and a great network is created, and if you do it right, you go home with 100 or more business cards and many great ideas to save tons of money for your company. It is your conference and it is all about you. We have been working hard to keep the world fleet flying. In the first place, I would like to tell you that I am privileged to be here with you sharing the passion for aviation. There are no borders to keep us apart. But more than ever, we need each other. The aviation environment is changing and there is continuous demand for safety, reliability, and profit. Over the last couple of years, we have also been experiencing new aircraft types entering the aerospace. If you have attended AMC conferences, you would remember that two years ago, the first question about a B787 component was submitted. Last year, we had many B787 questions. B787 become the real life. Airbus A350 entered the theater, too. We are actually in a transition period. We must keep the present fleet flying and learn many new things about new birds.

So let us talk a bit about the new cool stuff. We used to call aircraft aluminum bird. That term is no longer applicable for the aircraft. There is not much aluminum used to build a modern aircraft. The new materials are challenging because the aviation industry is not used to working with them. The new technology needs to be learned to be able to repair those new materials, to detect cracks, and to trust that such light parts can withstand the forces as good as materials we used to work with.

Many avionics components are interconnected by fiber optic cables. Other industries have been using fiber optic cables for at least 20 years. We are just entering the fiber optic environment and starting to learn cool new things. Is it a difficult technology? It shouldn't be! Thousands of cable guys are dealing with optic cables daily in our homes, buildings, actually everywhere. We should manage to deal with it in the aircraft but we have to learn how to use, test, and repair the optical electronics.

Data buses on present aircraft are too slow for new technologies. Therefore, a CAN bus is used on B787, A350 and A380. The CAN bus is what we call ARINC 825. Again, nothing new for the car industry, but new for aircraft. It is actually strange that BMWs have been using CAN bus for many years and we are starting to use it now. Again, no new technology, but new for us!

All the time, we are talking about level 3 repairs of printed circuit boards. We are experienced to repair the SMD printed circuit boards and the new stuff is already here. Yes, it can be done, but we have to learn and master the new technology called Ball Grid Array (BGA). It is common in our homes. All new computers have BGA chips; but we must master it in our workshops too.

Lead free soldering is not something new anymore. It is part of our life. If you do not know about it you should learn it as soon as possible. Do not wait to be surprised by lead free soldering.

Software is becoming a major subject in our life. You do not need the knowledge of writing the software code, but you need a lot of knowledge of how to manage the software. I am talking about buying, licensing, copying, transferring, loading, and backups. You need a good process in place to be able to manage hundreds of software parts which are flying in the modern aircraft.

Learning new technologies is not enough. To be part of maintenance force of B787, A350, or other new aircraft type, engineers should be aware that legal stuff is becoming crucial if you want to stay in business. It is important to have a contract but it is also very complex, especially if you are not a lawyer. Besides contracts, there are non-disclosure agreements, end-user agreements, designee letters, delegation letters, authorization letters, export licenses and ITAR licenses. For all those new subjects, they need engineers. Our life is going to change. Our activities will shift to other than engineering things. So here is the opening for

If you like it or not, we must change and adapt to the new world. If we resist change, our fate will be the same as the fate of the dinosaurs. And you all know what happened to dinosaurs 75 million years ago: because they could not adapt to a new situation, they vanished. For avionics shops, independent MROs, or airline owned MROs, it is a big challenge to stay in the business. The AMC | AEEC conference is the only place to get together and discuss technical problems of design of cool new stuff which AEEC is defining for us. If we stop this activity, every one of us will be alone and there is no way back. Only by working together, airlines, OEMs, and airframers at AMC | AEEC can make it possible to keep the conferences, exchange experience, and improve aviation.Hydro mechanical engineers are jealous of us because they do not have such a great conference and get together place. Do not forget that we, avionics engineers, are the right stuff. It is also a privilege to be a part of such an extraordinary group of people. Saying that, I would like to thank you once more for being here in Prague. Let us now hammer the start of the 66th AMC | AEEC conference.

Friday Night in Prague

The 2015 AMC conference in Prague finished on a Thursday. By that Friday night, everybody had left Prague to go home. I wondered if Prague would ever be the same after we left. Probably yes, but aviation will change a bit because in Prague we learned new things and we solved problems that have bothered us for some time.

We all discovered that we are at a beginning of a new aviation era. Some very familiar aircraft types are slowly changing into Classics. And the older classic aircraft are changing to Jurassic and cool new composite aircraft are entering the aviation theater. The transition will take some time (perhaps 5 years), and then we will talk about B747-8, A320-NEOS, B737-Max, A350, B787, and the B777-X. Some of the AMC engineers who were present in Prague will probably be retired before the first Level D check of those new aircraft types. Other engineers who were in Prague for the first or second time attending the AMC will endorse these new aircraft types with open arms and keep the avionics running for years to come.

In a couple of years, they will judge if we have done it right. They will either face problems or (what I certainly hope) they will face extremely reliable components and stable software. It is similar to a situation 25 years ago when we were used to fixing almost everything on B747-1/2/300, or the DC-10, or maybe the A310. After those birds, we got B747-400, MD11, A320 and a bit later, the A330. It was a step forward in technology and it was challenging (actually frightening). Suddenly, we had to think different. Computers were almost in every system and system integration was advanced to the stage that you could not just modify the systems without help from airframer. On the good side, data was easily accessible and although we were complaining, life was good.

Aviation technology has advanced quickly in the past 10 years. he present aviation environment is different than 25 years ago and although it looks complicated, it does not necessarily have to be complicated. Our industry is built by engineers native to earth, not by aliens! If they could develop, design, and build these new aircraft types, we should be able to maintain them. The technologies are different than before but we must manage to learn the cool new stuff.

Technology-wise, it is complicated but manageable. We all are smart enough to work with optical electronic, CAN busses, ball grid arrays, and loadable software parts. What bothers me is not the engineering stuff but the legal issues. Due to legal problems/issues, we will run into unwanted situations and into unwanted people. Yes, you are guessing, who are those people? Lawyers, of course!

I cannot start speaking about lawyers before telling you a lawyer joke. In a village, there was just one lawyer. He was poor and had no money and no food. Then the second lawyer moved into the village! After some time, the lawyers were the richest people in the village.

Now exchange the word village with avionics environment and it will sound like this:

In an avionics environment, there was just one lawyer. He was poor and had no money and no food. Than the second lawyer moved into the avionics environment! After some time, the lawyers were the richest people in the avionics environment.

Notice that I did not mention that the second story was a joke. Because it is not! It is deadly serious.

It bothers me that (besides that they become the richest people in the avionics environment) they have soiled the good relationship between operators and airlines. Many of us working on one side of the perceived line (say operator) like to be an operator and do not want or do not wish to change and become an OEM engineer.

Both sides want one thing: to have, perfect aircraft, and smooth operation. That has not changed in the past 100 years. So if it all were up to engineers, our life would be good.

Of course, reading this you would obviously ask: What is the problem? The answer is easy: Love!

Generally speaking, the love for aviation is disappearing. Everybody is demanding profit and pushing very hard to gain more market share. Airframers, airports, OEMs, banks, lessors, etc., are earning tons of money. The ones who are helping them are engineers and they are overworked and underpaid.

Everybody hates engineers but nobody can live without them. The good old times when we were proud of solving engineering problems are gone. Soon we may sell our souls and become advisors for our lawyers. We will tell them what can possibly hurt the pocket of our opponents or competitors.

We will tell lawyers which data is important and which is not. We will tell lawyers what to remove from data package to delay others to establish the capability of repairs. We will advise how to squeeze the opponents and how to frustrate them. Our souls will be sold to lawyers. That may be the future of avionics engineers.

The good old times when the Wright brothers were flying in Kitty Hawk, or when Lindberg was flying above the Atlantic or when Capt. Mission was flying B747-400 non-stop from London to Melbourne are gone forever. The love for aviation brought teamwork to past AMC conferences to share solutions.

Picture this romantic story (just 35 years ago):

The guys from the flight line called me complaining that the mechanical GMT clock in an F-27 was broken. I took a bicycle and tools to the tarmac and rode out to the airplane. I removed the clock and closed the hole with duct tape (yes, good old duct tape) to prevent cold air blowing in the face of the pilot. Then I biked to a shop called Hobby Clock in the nearest village to buy a spring for the clock. In the evening at home, I replaced the spring, checked the clock, and the next day I installed it in the aircraft. Everybody was happy and the costs were 1 USD (price of the spring). It was sound engineering judgment that the COTS spring was ok. The clock (mechanical type) was working well. It worked for the next 5 years until they retired the whole airplane. This is not possible any more. Not even a chance.

Is the aviation arena a safer place now? Yes, of course. It is not romantic but the industry is professional, regulated, and more expensive. All studies show that it is safer than ever before. In a way, we as engineers have done a great job. Maybe too good! One of my retired fellow engineers told me once: "A good engineer will make himself redundant."

In a way, she was right. We had some great systems engineers. Their systems were never in the top 50 systems of delays, there were never AOGs on their systems nor AD notes. At a certain point in time, management decided to scrap this position from the talent pool and made her redundant. All because her system was working perfectly and another engineer could do it simpler by bundling it together with his own job. But just one year later, the decision was reversed and the new engineering job was created because of too much work in that system. It was just that simple.

It is not romantic any more. However, I for one am confident that my pride in engineering work will never end. While pestered by these lawyers, I remain happy to keep the aircraft flying healthy.

Message from the AMC Chairman

KLM Royal Dutch Airlines
A Note from the AMC Chairman For annual Report 2015

Every year I say to myself: the aviation industry is more complicated and more dynamic than ever. When I say complicated, I don't mean only technically complicated, but also full of complicated interactions between airframers, operators, MRO's, brokers, lawyers, and engineers. Every year I have the feeling that it can't be more complicated and more dynamic and that we are at the top of the curve. Every year I hope that we will start to descend and get more stability, hoping that everything will calm down. But...every year I notice that it can be more complicated, more regulated, more dynamic, more challenging, and more intensive. Obviously we can't change it overnight and not even over a long period. We have to deal with it. We had periods of merging of operators. We had periods of creating MROs and making them independent of operators. Also, we were running into intellectual properties wars, rather than lowering costs, introducing new aircraft types, phasing out present fleets and suffering of market demands to lower the ticket prices.

By the way, not long ago, fuel prices were extremely high and now they are extremely low. Those developments and interactions will obviously not stop. Our only choice is to deal with it. Therefore, we must be willing and able to change and be flexible no matter what. 75 million years ago there were species on earth who were controlling the environment. Yes, they were called dinosaurs. They were strong, dominating, and had plenty of food. They were kings and queens of the earth. They could go wherever they wanted and they could do whatever they desired. And then a big chunk of stone crashed somewhere on earth, caused big climate change. Dinosaurs could not adapt to the new circumstances and died. Other inferior species survived because they were able to adapt to the new climate. They were more flexible, perhaps more willing, and of course more able. They started to change and evolve into us.

Why I am telling you all of that? Because there is some similarity with dinosaurs! They say that history is repeating itself. That is absolutely true. In aviation, we are in the process of change. New aircraft (B787 and A350) are changing aviation. If we copy dinosaurs, probably we will have the same fate. To survive, we must act differently. The secret of surviving is to be flexible, to dare, and to be willing to change. If we say the crisis will blow away, we are wrong. We have to deal with every little detail of the situation we are facing. We have to be open to changes, to learn new techniques, to study and educate ourselves to learn new things and "boldly go where no man has gone before." Sometimes we will have to offer our time and patience. Sometimes we will even not get a sufficient reward, but we have to focus on our ultimate and noble target: never compromise safety and make all decisions with healthy engineering brain and follow your heart.

Most of aviation boys and girls share an endless passion for airplanes. Sometimes, engineers can be disappointed by not getting the cooperation from OEM's, or having difficulties not with technical stuff but with procedures, contracts and regulations. The easiest solution is to step over to another industry. The guys from dot.com love us. People from oil platforms and oil business love us and in most cases, they pay better. They love us because we are dedicated to our projects. They love us because they know that we know how to deal with procedures, continuous improvements, and safety. They know that we are good. Therefore, I have full, complete, and immeasurable respect for hard working engineers in our beautiful industry.

Besides all hard work at our companies, a bunch of individual engineers dedicated to aviation are organizing and attending the meetings, setting standards, improving each other's know-how, and making sure that thousands of aircraft are flying every day, transporting millions of passengers and bringing tons of packages to the right destination. Only cooperation, flexibility, dedication, and hard work can make it possible. The bottom line is that we all have to do a lot of work to comfort our customers who have no mercy and who are only demanding lower ticket prices. This is the reality and we have to deal with it. I guess that on December 17, 1903, the Wright brothers didn't have any idea what they caused. They were actually the cause that other industries will downsize and aviation will grow. Again, similarity with the dinosaurs! Before the Wrights, big (dinosaur) ships were transporting people from continent to continent. Then, came the blow in the Kitty Hawk! And the new species entered the arena, the airplanes. The bicycle makers started the big change, causing the founding of ARINC just 25 years later. The rest is history but ARINC Industry Activities is here all the time helping us all to keep changing and surviving. So endorse the changes and be part of it and don't forget the fate of dinosaurs.

Before Something Nice Happens—Chaos

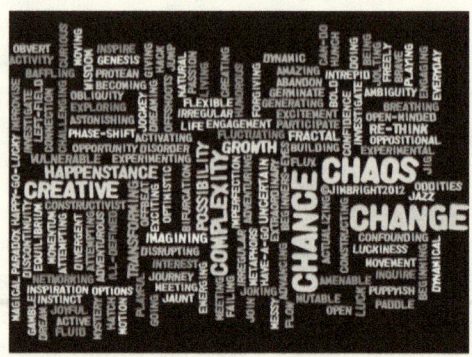

Everything is relative: that is what Einstein taught us. Even in aviation, the concept of relativity can apply to the industry's stability. In the short term, it looks like we are in a very unstable aviation environment. I will give you an example: new aircraft like the Airbus A350 and the Boeing B787. In this case, lawyers and economists have designed the new business models. We were a generation of engineers who grew up with jet aircraft and we did not have much feeling for the "Jurassic" propeller-driven airships like the Lockheed Constellation or Electra. We did not care about those aircraft because we were young and we wanted to be part of the jet engine workforce. The revolution lasted at least 10 years, and when all propeller aircraft were replaced by the new jets, we built a new environment to last the next 40 to 50 years. If you look back at the transition period now, you do not remember that it took 10 years to adjust. But at that time, the business model was also different. Propeller aircraft did not last as long as jets and there was a lot more maintenance. We are in a similar situation now. We are changing from aluminum aircraft to composite. These new aircraft are lighter, more reliable, consume less fuel, and are full of software.

The business model has changed and OEMs are taking over the aftermarket. Lawyers are playing a more prominent role. There are some that are happy, but not everybody. Everything is much more difficult to accomplish. The strategy is to adjust or perish like the Jurassic dinosaurs. History is repeating itself, but in a different form. 10 years from now, we will think differently and we will be used to a new world. For now, it is chaos. We must learn new technologies and cope with regulation, software management, lawyers, and contracts. Looking back, the chaos started 4 or 5 years ago. Remember the times when you could buy a DC-3 (known as Dakota) by signing a one page contract? Those were the days, my friend. Saying that, I have to come back to the following story:

A surgeon, an engineer, and a lawyer are having a heated discussion in the pub concerning which of their professions is actually the oldest profession. The surgeon says: "Surgery IS the oldest profession. God took a rib from Adam to create Eve and you can't go back further than that." The engineer says: "Hold on! In fact, God was the first engineer when he created the world out of chaos in 7 days, and you can't go back any further than THAT!" The lawyer smiles and says: "Gentlemen, Gentlemen...who do you think created the CHAOS??!!"

This is our new benchmark: the chaos at present. I estimate that we will need at least 5 more years to stabilize the situation and get an overview. We will then have composite aircraft with stable software, reliable LRUs, and low-cost operation. There will be a few big MROs, a lot of OEM shops, and many small shops which will maintain low cost, simple to fix LRUs. Lawyers and legal departments will work full-swing in this new reality. Engineers are constantly under pressure to design flawlessly and deliver on time. However, it can take weeks or months to produce 60 pages of a contract. Then, after many months of discussing it, it is signed. The next day, the manager will call the engineer to ask: "Is the test already in? What? 9 months lead time? Are you insane? We signed a contract. Ok, dear engineer, this is definitely not what we want. Call them and ask to deliver it next week. Tell them our production department is waiting. We need it now! We are a customer!" This chaotic situation will not last forever. After these troublesome years resolve themselves, we can relax and enjoy our retirement. The new workforce will take over just like we took over from the propeller guys. Keep smiling.

On behalf of the AMC Steering Group, we look forward to seeing you for the AMC | AEEC in Atlanta, April 2016, hosted by Delta Air Lines. This will be the 67th annual meeting of this conference. The only constant factor in aviation industry is change. The period since the last AMC was no exception. Many B787s and A380s are flying. A350 is entering the theater too. Life will never be the same because new birds are different. We are already used to idea that MD-11 and B747-400 are leaving or going to leave the skies. We are looking forward to greet the A320-NEO and B777X in the near future. Airframers and OEMs are doing a great job to deliver reliable and safe aircraft, which will spend the majority of useful life in hands of operators. Actually they delivered a cash generator to airlines but it is not that easy.

The difficult part is to make sure that airlines effectively use those money generators. In another words, we are at the turning point. New aircraft, new components, new technologies, new contracts, new playing field. Everybody must change. The most terrible move you can do is to stick in the old pattern and do nothing, thinking that the wind of change will blow over. Unfortunately, the wind of change keeps blowing and nobody can stop. Therefore, respect new rules, business cases, and environment and adapt. The ones who are not able or not willing to adapt will vanish just like dinosaurs vanished 75 million years ago. They will go and will be forgotten. One factor which can point the way is the incredible brain power which can only be experienced at AMC conference. This is not just a conference: it is THE Conference. You can't find a better place and you can't, definitely can't, find a place with more knowledge about modern aviation.

The professionals of SAE-ITC improved Industry Activities and made it stronger, which is good! Airlines and OEMs are there to elaborate, discuss problems, provide solutions, and give unbiased advice. The only thing you should do is to come over to Atlanta and explain what is bothering you. Between 650 professionals, there is always at least one who can help you. Even if you didn't submit your question in advance, just go to an airframer representative, supplier, or airline engineer and ask. People will help you because they care. They want to help. They are willing and they are able to help. Just ask. Today they will help you, tomorrow you might help them. That is the spirit of the conference.

At AMC, we also care that everybody is educated; therefore, we organize seminars. You could say that seminars are customized just for you, me, and a broad audience. It is not a coincidence that we issue a questionnaire at every conference. Based on those inputs, we carefully select seminars which are of most interest. This time there is no exception. On your request, we attracted the best speakers in the industry to provide unbiased, honest, and high quality information just for you. The

106

information is worth money and if you use it intelligently, it can generate a lot of cash for your company. We sincerely appreciate the continued support from our Member Organizations and Corporate Sponsors of Industry Activities that enable us to conduct this one of a kind global event for the benefit of all our attendees.

It is our duty as operators, OEMs, and airframe manufacturers to spread the word about who we are and what we do. As a supplier or airframer, please remember that being an AAI member is an important contribution to AMC, but if your organization has experienced increasing success as a result of participating in AMC, please consider also becoming a Corporate Sponsor of Industry Activities. By growing our memberships and sponsorships, we can ensure the future of AMC as your primary source for constructively resolving technical issues, providing information exchange, and networking with business partners you might only see this time of year. OEMs, of course, have their justification for attending the AMC conference. It is a unique opportunity to meet and greet more than 45 airlines. It is a dramatic cost saving to go to AMC instead of traveling to 45 or more destinations and spend many days on a trip. If they plan it well, they can meet 45 airlines in just 4 days. Airlines can meet 150 OEMs in the same period of time instead of having 150 meetings spread across typically a six month period. Actually, everybody wins. Which is excellent.

Also, I would like to express my great respect and appreciation for Delta airlines which is hosting the conference. Delta Air Lines' engineers know the benefits and know the challenges of the aviation industry and besides all hard work to make their airline one of the best and most respected on the planet, they also spend their additional time and effort to host the conference. I would like to say big thank you for all good care: THANK YOU DELTA AIR LINES. Finally, I would like to take this opportunity to thank the ARINC IA staff and my steering group associates for the additional effort to make the AMC happen for the 67th consecutive year.

My steering committee members are delivering additional effort on the top of their normal airline activities to help organize and improve the conference. Their companies are convinced that there is a good payoff for all the effort because they realize that the more you put into ARINC IA, the more you benefit. Thank you all for being so cooperative. Saying that, I would like to remind you that Atlanta is the place and April 24, 2016, is the time. I will be there and make sure that you will be there too. Atlanta will be the hottest place in the universe, make sure that you are there too.

Packing for Atlanta

It is almost time. The clock is ticking and countdown has started. Just before 0830 on April 24, 2016, Smitty will fire up the AMC tune, Sam will ring the cowbell, and the 67th AMC | AEEC conference will begin. This time it is hosted by our dear friends from Delta Air Lines. Once more, we will revive the tradition. People from all over the world will join in the meeting room, excited to solve problems, earn cash, save money, and learn new things. AMC and AEEC engineers will meet again. This is how the AMC conference works! Step one is to submit the questions in advance. Everyone is asked to submit questions, but there is one simple rule: discuss your problems with airframers and suppliers all year long. If they are not helping or are unwilling to help, tell them that you will submit the question to the AMC conference. If that does not help and they are still not cooperative, go ahead and submit the question. Each year we collect approximately 200 to 250 questions, which is good. Do not hesitate to ask, even if you think: "Oh man, this is a stupid question!" Be assured that the only stupid questions are those that are not asked. The questions are published in February and there is a window of time for suppliers to prepare their answers. While somebody is working on your questions, your task is to read the questions of others. You might learn new things, find that you have a similar problem, or you may discover that you have a solution for a problem submitted by another. This is the information that we need from you.

All those questions and answers will be handled during our conference. Open your ears and contribute to these discussions. Only together can we create success. At the end of each day, there is a seminar containing topics engineers have selected for the conference. You will be amazed by quality of the topics and presenters. Only the best minds in the industry present at the AMC. It is an honor and a noble task to share this knowledge with others. People might ask: why you are doing all of this at a conference? 2015 was the safest year in the aviation! That's why! Besides the official part of the conference, there are some interesting collateral items. There is the AAI reception on Tuesday evening. This is a socalled "MUST ATTEND" event. For some, it is the most important part of the conference. Suppliers show off their new products, like new DER repairs or new PMA parts. This is the place to be if you want to save cash for your company. This is the hottest evening event. Each day in the evening hours, you can visit the hospitality suites and talk to suppliers. There is plenty of time for fun but also plenty of time for business. Take your chance and focus. This is a once-a-year event that should not be missed.

And just to remind you! ARINC Industry Activities (IA) is also an environment that produces aviation standards. All year long, small working groups loaded with specialists are meeting and designing standards. This year, you might also attend some of these meetings and contribute. These standards can save you money, and if you contribute, there is an opportunity to steer them to fit your needs. Otherwise, decisions will be made without you and the only thing that you can do is to follow. You might not like it, but others have decided for you and you did not even have a vote. So believe me: it is always better to lead than to follow. If I have a choice, I will always choose to lead knowing that a leader is one who knows the way, goes the way, and shows the way. Finally, some thoughts about other important things! The AMC | AEEC conference does not just happen. It is organized by a small group of people called the AMC Steering Group. This group manages an event full of activities, like the hotel, 700 conference attendees, and making sure everything runs like a well-oiled machine. Also, these same individuals are making sure that we stay within the budget and have the greatest experience you can have, now for the 67th time. I am more than honored and privileged to be a chairman of it all.

And the Octopus Takes the Cheese!

With the introduction of the new generation of aircraft come new business model for Original Equipment Manufacturers (OEMs). In order to play a role and get on board the aircraft, OEMs must invest resources and effort in helping airframers design the aircraft systems. The very moment that philosophy took place, airframers scored a deal; but, at the same time, they killed the competition. We all remember the time of Terrain alert Collision Avoidance Systems (TCAS). At that time, there was more than one OEM developing aircraft components. The competition between Honeywell, Rockwell Collins, and Allied Signal was intense, but the end users (airlines) were very happy because they could negotiate the price of their Line Replaceable Units (LRUs). The fate was in the hands of operators. In the case of the modern B787 or A350, you cannot do that — there is a single source and that is it. Is the aircraft cheaper? I don't think so! But the OEMs have now captured the aftermarket, which was traditionally in the hands of airlines because we all were sleeping. Most big airlines had excellent and huge avionics shops, a lot of new technology, and a lot of capabilities. It is now a difficult task to compete with the new business system. The OEMs had to plan their strategy carefully and be very patient. MROs are now complaining, because they see and feel that they are losing ground in this cutthroat competition. This is what we call the survival of the fittest. MROs are now in the grip of the octopus. The tentacles are strong. Each tentacle can strangle the MRO shop if they wish to do it. But they do not want to because the MRO's are still bringing cash to octopus. Let's learn more about these octopus tentacles.

Tentacle number one is the diminishing content and accessibility of repair manuals. Although that reduction of content is not rampant, OEMs have still moved content to different chapters or even different books. Or they remove it from Component Maintenance Manual (CMM) and insert it in the Technical Support and Data Package (TSDP). They are not obligated to provide TSDP's per PSAA as they are for CMM's. To get the TSDP you must sign an NDA. In that case, you are in hands of lawyers. And once when you are in hands of lawyers: well that is another story. An especially long story. They (not lawyers but OEMs) also have the power to issue revisions almost every day. By the time the MRO incorporates the revision, the next revision is already on the shelf. That is exhausting for the engineers responsible for CMM revisions. Third party work is absolutely forbidden with those CMM's unless you arrange it legally via delegation letters.

Now, meet **Tentacle number two**: High annual price escalation for LRUs and parts. If it is about Seller Furnished Equipment (SFE), then there is a limitation imposed by a Product Support Agreement (PSA). For Buyer Furnished Equipment (BFE), it depends on the quality of your own contract. Again, there are additional costs for MROs but now days MRO's are lucky that there are no many BFE components. On the other hand, if you sign the contract for SFE components, be careful because lawyers will insert small print: By signing the contract you declare that the PSA agreement is no longer valid. You are dead!

Tentacle number three is the extreme long delivery performance of the OEMs. Lead times are long for everything. If you buy the new stuff, expect lead times of 240 to 360 days. It is the same with piece parts, and it is not getting better.

Tentacle number four: Upgrading parts or making parts obsolete. This is a difficult one. Although there is an obsolescence plan required for each OEM, many times they do not comply. They always have a number of excuses why they cannot comply. If OEMs foresee a positive outcome, they declare a part obsolete and try to sell a new redesigned LRU, which is better (of course), more reliable (of course) and more expensive (naturally). OEMs will oppose PMA parts or DER repairs. PMA parts are nowadays better than ever, but they are getting bad publicity. People who are not engineers can easily decide to ban PMA parts and the little bit of competition against the OEM will be extinguished. Lessors are working together with suppliers and helping to ban the PMA parts. Again beware of lawyers who will insert a sentence like: MROs will get a discount of 0.05% on all parts but must promise to never PMA parts or DER repairs.

Tentacle number five is ensuring limited availability of alternative parts and repairs. This tentacle is difficult to manage. You know that repair is possible but the CMM is forcing you to replace the part. If you want to do a repair, which is not described in the CMM, you must spend a lot of effort and money to get the approval for it. And after all that effort, your customer might not accept the repair or alternative PMA part. So you are cornered in your misery. You will notice that OEM's are doing repairs using their own documents but they will never give them to you even if you want to buy them.

Then there is **the sixth tentacle**: limited access to tooling, education and technical support. If you want to buy tooling, it is typically not offered, or affordable. If you want a drawing, it is proprietary. If you want to buy a drawing, the price is not available. If you fill out the form on their web portal and request training you will get a computer generated email saying that the job is in process. After several updates later, they will give you the wrong answer anyhow. Eventually they put the statement in CMM: Only Trained people of OEM are able to do the repairs. Send all repairs to the OEM along with a lot of your airline's money.

The seventh tentacle is all about creating barriers for third party work. OEM will typically insist on intellectual property agreement, end-users agreements and non-disclosure agreements. License fees are always imposed for such third party work in addition. In the old days, you could buy parts for your needs and for your customers. Now you cannot. They will give you higher price or force you to enter the end users agreement which entails endless discussion between your lawyers and the OEM's lawyers. At the end of the discussion they will ask royalties for each serial number you repaired for your customer.

Tentacle number eight prevents increase of your own revenue by pushing for additional business or, in other words, the OEMs increase the Turn Around Time (TAT) of repairs of LRUs and SRUs to force you to buy additional spares to fill the pipeline. The OEMs are also able to keep discussion about your quote or price of parts going just long enough that you find yourself regularly in an AOG situation. Eventually, you will decide to buy one more spare LRU just to manage the risk of AOGs. But that is exactly what they want: to sell you additional spares. If you need it fast, you pay a higher price.

So now you know everything about the octopus. It is time to meet the cheese. The story is known about the farmer in the dell. It says:

> **"The farmer takes a wife.**
> **The wife takes the child.**
> **The child takes the nurse.**
> **The nurse takes the cow.**
> **The cow takes the dog.**
> **The dog takes the cat.**
> **The cat takes the mouse.**
> **The mouse takes the cheese.**
> **The cheese stands alone.**
> **Heigh-ho, the derry-o, the cheese stands alone..."**

And YOU are the cheese my dear engineer! The cheese stands alone. YOU, cheese, are at the end of the food chain. Everybody expects that you will fight all tentacles of the octopus. You know the problem. You feel it. All other layers in the food chain need your knowledge and your information to tell you later: I have fixed your problem! Go ahead and develop the capabilities. Sometimes I regret that I am an engineer and not a lawyer. Lawyers are the kings and the queens of the company, and they do not mind if they work for OEMs or MROs. They make the policy, establish the tactics and make the rules, allocating points and commas.

114

They have all the fun and everybody respects them. We engineers just suffer, we are the cheese. We are a pain in the neck because we always complain that we need more data and more information.

Each individual tentacle is capable of strangling you, just not very quickly. The octopus is just playing the game and seeing just how long you can suffer. If you have survived the stranglehold of the octopus so far, then you are on a good path. Shortly you will conclude that there is a partnership required if you want to survive, perhaps even to relax the stranglehold. When you think that the solution is the partnership with another MRO to create the greater magnitude and better negotiation position or the partnership with OEM to create both a win-win situation, you will realize that you have become somebody who suffers and generates money for the octopus. If you partner with Octopus, it is even more terrible because after some time, the octopus will start to strangle you but you cannot escape because breaking the partnership costs more but than operating alone. It is not possible anymore, because the octopus is much stronger and you are just the cheese.

TORONTO: In Memoriam Axel Muller

by Marijan Jozic, AMC Chairman

Thank you for the years we shared
Thank you for the way you cared
Rest in peace, dear engineer and friend.
The memories will never end.

Axel dedicated his life and career to aviation. Axel spent more than 45 years at Lufthansa. Besides a successful career at Lufthansa, he was extremely active in ARINC Industry Activities. I met him for the first time back in the early nineties, in Hamburg, when he was in charge of the Lufthansa avionics shop. He provided a group of us a tour through the components shops. It was very impressive! He knew every single piece of equipment and could tell a story about it. It was very unusual that a high shop manager knew so many details. His technical knowledge was unprecedented. Years after that, I ran into Axel at the AMC in Baltimore. That was in 1999. It was my first AMC but not his. It has been my privilege to share all these years with such remarkable man. If you look from the sidelines you can only conclude that Axel was a driving force in generating many of the ARINC standards that now benefit AMC.

Axel was veryknowledgeable about how the aerospace business operates, how industry activity operates, and how to keep it under control. In all the years we have learned a lot from Axel because he was not keeping the knowledge for himself. He was a member of the AMC Steering Group for many years and served as AMC Chairman for 4 years. Axel served as Chairman for numerous ARINC standards activities. Also, his ideas about future of our industry made us aware that we cannot operate alone and that we need good dialog to be able to operate and cooperate. Axel was awarded with the Volare Award in 1996 and Man of the Decade Award in 2009. After Axel's official retirement from Lufthansa, Axel was still participating in many of the ARINC Industry Activities Standards Activities (ARINC 668, 602B, 663, and 674 to name a few). Axel also planted a seed in Tulsa. He actually started a brand new Lufthansa shop in Tulsa. During the 2008 AMC in Tulsa, I had the opportunity to visit his shop. It was a shop with just 3 employees: Axel and two more. Axel was not only a manager, but also a lead technician, certifying stuff, contract manager, human resources, and purchasing and sales guy, which is very unusual. But that is exactly the way real engineers operate. Now 6 years later there are more than 80 techs working in the component shop that Axel initiated. Great job!

Not many people know that Axel was also a musician. Several times he told me that if he did not become an engineer he would have been a professional musician. Many times at AMC we had the opportunity to sing and play in hospitality suites. We enjoyed when he was playing trumpet or piano. Thank you Axel for entertainment! Not long ago we got bad news from Germany. Axel was ill and would not get better. He was in Orlando last year at his beloved AMC. That was his last conference. We saw that he was ill, but we all were hoping that he would stay with us for many years to come. Unfortunately, the best possible medical care did not help and he passed away just 2 weeks before the Toronto conference. It was our great privilege to know Axel in person and to be his friend and colleague. It was my privilege to know him and to learn from his experience. We will miss him a lot but he will always stay in our memories.

Thank you for the years we shared.
Thank you for the way you cared.
Rest in peace dear engineer and friend.
The memories will never end.

Opening Speach PRAGUE 2015

Marijan Jozic, KLM Royal Dutch Airlines
AMC Chairman 2015

Good morning my fellow engineers!

I hope that you all found some time to see the beautiful city of Prague. It is one of the most beautiful cities on the old continent. It used to be one of 4 major cities in Habsburg monarchy, together with Vienna, Budapest, and Zagreb. Each of those cities is now in a different country.On behalf of the AMC Steering Group, welcome to the 66th AMC | AEEC conference! Welcome to Prague!

The most charming part of the AMC | AEEC conference is that it is a magnificent formula. Just to remind you: 600 professionals come from all over the globe, meet for 4 days, and solve each other's problems. Many new ideas pop up, new friendships, and a great network is created, and if you do it right, you go home with 100 or more business cards and many great ideas to save tons of money for your company. It is your conference and it is all about you. We have been working hard to keep the world fleet flying. In the first place, I would like to tell you that I am privileged to be here with you sharing the passion for aviation. There are no borders to keep us apart. But more than ever, we need each other. The aviation environment is changing and there is continuous demand for safety, reliability, and profit.

Over the last couple of years, we have also been experiencing new aircraft types entering the aerospace. If you have attended AMC conferences, you would remember that two years ago, the first question about a B787 component was submitted. Last year, we had many B787 questions. B787 become the real life. Airbus A350 entered the theater, too. We are actually in a transition period. We must keep the present fleet flying and learn many new things about new birds. So let us talk a bit about the new cool stuff.

- We used to call aircraft aluminum bird. That term is no longer applicable for the aircraft. There is not much aluminum used to build a modern aircraft. The new materials are challenging because the aviation industry is not used to working with them. The new technology needs to be learned to be able to repair those new materials, to detect cracks, and to trust that such light parts can withstand the forces as good as materials we used to work with.

- Many avionics components are interconnected by fiber optic cables. Other industries have been using fiber optic cables for at least 20 years. We are just entering the fiber optic environment and starting to learn cool new things. Is it a difficult technology? It shouldn't be! Thousands of cable guys are dealing with optic cables daily in our homes, buildings, actually everywhere. We should manage to deal with it in the aircraft but we have to learn how to use, test, and repair the optical electronics.

- Data buses on present aircraft are too slow for new technologies. Therefore, a CAN bus is used on B787, A350 and A380. The CAN bus is what we call ARINC 825. Again, nothing new for the car industry, but new for aircraft. It is actually strange that BMWs have been using CAN bus for many years and we are starting to use it now. Again, no new technology, but new for us!

- All the time, we are talking about level 3 repairs of printed circuit boards. We are experienced to repair the SMD printed circuit boards and the new stuff is already here. Yes, it can be done, but we have to learn and master the new technology alled Ball Grid Array (BGA). It is common in our homes. All new computers have BGA chips; but we must master it in our workshops too.

- Lead free soldering is not something new anymore. It is part of our life. If you do not know about it you should learn it as soon as possible. Do not wait to be surprised by lead free soldering.

- Software is becoming a major subject in our life. You do not need the knowledge of writing the software code, but you need a lot of knowledge of how to manage the software. I am talking about buying, licensing, copying, transferring, loading, and backups. You need a good process in place to be able to manage hundreds of software parts which are flying in the modern aircraft.

- Learning new technologies is not enough. To be part of maintenance force of B787, A350, or other new aircraft type, engineers should be aware that legal stuff is becoming crucial if you want to stay in business. It is important to have a contract but it is also very complex, especially if you are not a lawyer. Besides contracts, there are non-disclosure agreements, end-user agreements, designee letters, delegation letters, authorization letters, export licenses and ITAR licenses. For all those new subjects, they need engineers. Our life is going to change. Our activities will shift to other than engineering things. So here is the opening for us engineers. If you like it or not, we must change and adapt to the new world. If we resist change, our fate will be the same as the fate of the dinosaurs. And you all know what happened to dinosaurs 75 million years ago: because they could not adapt to a new situation, they vanished. For avionics shops, independent MROs, or airline owned MROs, it is a big challenge to stay in the business. The AMC | AEEC conference is the only place to get together and discuss technical problems of design of cool new stuff which AEEC is defining for us. If we stop this activity, every one of us will be alone and there is no way back.

Only by working together, airlines, OEMs, and airframers at AMC | AEEC can make it possible to keep the conferences, exchange experience, and improve aviation. Hydro mechanical engineers are jealous of us because they do not have such a great conference and get together place. Do not forget that we, avionics engineers, are the right stuff. It is also a privilege to be a part of such an extraordinary group of people. Saying that, I would like to thank you once more for being here in Prague. Let us now hammer the start of the 66th AMC conference.

In the Grip of the Octopus

Written for Avionics Magazine (promised to avionics Magazine at Toronto AMC conference)

With the introduction of the Boeing 787 and the A350 come new business models for Original Equipment Manufacturers (OEMs). In order to play a major role, or just any role at all in aircraft systems, OEMs must invest many resources and much effort in helping Boeing and Airbus design the aircraft systems through all the high costs of development. Under all the pressure, airframers argued that it would be much cheaper and easier if they only had to certify one system. The very moment that philosophy took root, airframers scored a deal; but, at the same time, they also killed the competition. We all remember the time of Terrain alert Collision Avoidance Systems (TCAS). At that time, there was more than one OEM developing aircraft components. The competition between Honeywell, Rockwell Collins and Allied Signal was intense, but the end users (airlines) were very happy because they could negotiate good price of their Line Replaceable Units (LRUs). Now, in the case of the modern 787, you can't do that — there is a single source and that's it. Is the aircraft cheaper? Not really, but by doing that and sharing development costs, airframers created a new business environment. The OEMs now have to capture the aftermarket, which was traditionally in the hands of airlines, which were able to make some money doing maintenance.

Avionics Shops

Most big airlines had excellent and huge avionics shops, a lot of new technology, and a lot of outstanding technicians. It is a difficult task to compete with that system, which had been running for decades, and OEM's had to plan carefully in advance and be very patient. But now, MROs are complaining, because they see and feel that they are losing ground in the cutthroat competition that is the survival of the fittest. While they once wanted the challenge, MROs are now recognizing the cost and find themselves in the grip of the octopus.

The tentacles are strong and each one of them can strangle the MRO shop. Let me introduce you to the tentacles!

Tentacle number one is the diminishing content and accessibility of repair manuals. Although that reduction of content is not rampant, still OEMs have moved content to different chapters or even different books, which is very confusing. They also have the power to issue revisions almost every day. By the time the MRO incorporates the revision, the next revision is already on the shelf. But it is easy for the OEM to keep revising manuals, because the digital nature of distribution means they no longer have to print it and distribute physically. Just revise something, anything, and place it on the web. If you are the airline, you are allowed to use the manual only for your own fleet. Third party work is absolutely forbidden unless you arrange it legally via delegation letters. And it is not becoming easier. That means new costs, which are added to the costs of the MRO product, making that MRO less competitive.

Now, meet tentacle number two: high annual price escalation for LRUs and parts. If it is about SFE, then there is a limitation imposed by a Product Support Agreement (PSA). For BFI, it depends on the quality of your own contract. Again, it means, among other things, additional costs for MROs.

Tentacle number three is the poor delivery performance of the OEMs. It looks like a game and sometimes appears as a coincidence, but to me there appears to be a pattern. Lead times are long for everything. If you buy the new stuff, expect lead times of 240 to 360 days. It is the same with parts. And it is not getting better. Yes, tentacle three is the dramatic complication and additional costs for the MRO.

Tentacle number four: upgrading parts or making parts obsolete; this is a difficult one. Although there is an obsolescence plan required for each OEM, many times they don't comply. If OEMs need more revenue, they declare a part obsolete and try to sell a new redesigned LRU, which is better (of course), more reliable (of course) and more expensive (naturally). The game can be played indefinitely. OEMs will of course oppose PMA parts or DER repairs. PMA parts are nowadays better than ever, but they are getting bad publicity.

People who are not an engineer can easy decide to ban PMA parts and the little bit of competition against OEM will be extinguished. Nevertheless, MRO costs will again get higher as they can only use marked-up original parts.

Then there is the fifth tentacle: limited access to tooling, education and technical support. If you want to buy tooling, it is typically not put on the table. If you want a drawing, it is proprietary, and if you want to buy a drawing, they don't have a price available. On the other hand, the training is not scheduled; instead, it is on request and there is always some problem, because you are the only one asking for training. Getting technical support is also a problem because you can't get real people on the telephone line. You must send them an email, which will be responded to by a computer generated email or maybe even the email address will be unknown. If you fill out the form on their web portal, you will get a computer generated email saying that the job is in process. After several updates later, they will give you the wrong answer anyhow.

Tentacle number six is ensuring limited availability of alternative parts and repairs. This tentacle is difficult to manage. You know that there is a repair possible but CMM (Component Maintenance Manual) is forcing you to replace the part. If you want to do a repair, which is not described in the CMM, you must spend a lot of effort and money to get the approval for it. And after all that effort, your customer might not accept the repair or alternative PMA part. So you are cornered in your misery.

Tentacle number seven prevents increase of your own revenue by pushing for additional business or, in other words, OEMs increase the Turn Around Time (TAT) of repairs of LRUs and SRUs to force you to buy additional spares to fill the pipeline. The OEMs are also able to keep discussion about your quote or price of parts going just long enough that you find yourself regularly in an AOG situation. At a certain moment, you will decide to buy one more spare, just to manage the risk of AOG. But that is exactly what they want: to sell you additional spares.

The eight tentacle is all about creating barriers for third party work on intellectual property agreements, end users agreements and non disclosure agreements. License fees are always imposed for such third party work in addition. In the old days, you could buy hardware parts for your needs and for your customers. I can imagine that the OEM asks a fee now for sending a copy of software. Suddenly you need to buy a screw to repair the unit of your customer (not software, but a screw, thus hardware). It is not enough that you are charged for the screw, but they are asking the fee because it is the unit of your customer. So, they s---w you twice! Isn't that strange? This is brand new for airlines and MROs, which are not used to do so much legal stuff.

Sometimes I regret that I am an engineer and not a lawyer. Lawyers are nowadays the kings and the queens of the company, and they don't mind if they work for OEMs or MROs. They are making the policy, establishing the tactics and making the rules. Actually, they have all the fun. We engineers just suffer.

Every individual tentacle is capable of strangling you, just not very quickly. The octopus is just playing the game and seeing just how long you can suffer. If you have survived the stranglehold of the octopus so far, then you are on a good path. Shortly you will conclude that there is a partnership required if you want to survive, and perhaps even relax the stranglehold. When you think that the solution is the partnership with another MRO to create the greater magnitude and better negotiation position or the partnership with OEM to create both win situation, you will realize that you have become somebody who just suffers and generate money for the octopus.

To operate alone? It is not possible anymore, because the octopus is much stronger.

It is always better to lead than to follow

People might ask: why you are doing all of this at a conference? 2015 was the safest year in the aviation! That's why! Besides the official part of the conference, there are some interesting collateral items. There is the AAI reception on Tuesday evening. This is a socalled "MUST ATTEND" event. For some, it is the most important part of the conference. Suppliers show off their new products, like new DER repairs or new PMA parts. This is the place to be if you want to save cash for your company. This is the hottest evening event. Each day in the evening hours, you can visit the hospitality suites and talk to suppliers. There is plenty of time for fun but also plenty of time for business. Take your chance and focus. This is a once-a-year event that should not be missed. And just to remind you! ARINC Industry Activities (IA) is also an environment that produces aviation standards. All year long, small working groups loaded with specialists are meeting and designing standards.

This year, you might also attend some of these meetings and contribute. These standards can save you money, and if you contribute, there is an opportunity to steer them to fit your needs. Otherwise, decisions will be made without you and the only thing that you can do is to follow. You might not like it, but others have decided for you and you did not even have a vote. So believe me: it is always better to lead than to follow. If I have a choice, I will always choose to lead knowing that a leader is one who knows the way, goes the way, and shows the way.

Finally, some thoughts about other important things! The AMC | AEEC conference does not just happen. It is organized by a small group of people called the AMC Steering Group. This group manages an event full of activities, like the hotel, 700 conference attendees, and making sure everything runs like a well-oiled machine. Also, these same individuals are making sure that we stay within the budget and have the greatest experience you can have, now for the 67th time. I am more than honored and privileged to be a chairman of it all. Be there too.

FOM: Future Obsolescance Management

Interview with Marijan Jozic! AMC Chairman and Development Manager at KLM Royal Dutch Airlines

Please tell us about yourself.

My professional and personal interests are quite diverse. When I am not busy in the field of aviation, I am involved in many other things:

- Writing books and articles about aviation
- Writing short articles on various subjects for a Croatian Internet site
- Gardening
- Repairing things in and around the house
- Sea fishing
- Reading science fiction books and biographies

What is your current professional position, and what career path led you to it?

I started my professional career as a test equipment calibration engineer. Following that, I was a flight guidance systems engineer for 14 years, then a senior avionics engineer, and eventually an EASA verification engineer. The next steps in my career were a move to the business project office at KLM Royal Dutch Airlines as a modification manager for the wide-body fleet and then to a position as 747-400 account manager. Currently I am a development manager at KLM, leading a team of 27 development and production engineers who are building new capabilities and making repairs in the KLM avionics shop. Parallel to that career, and separately from KLM, I am on the board of project managers for organising the annual aviation knowledge and career day at the Technical University of Amsterdam. In addition, for the past four years I have served as a member of an advisory committee for the Dutch Aerospace Laboratory. Also, I have been a member of the Avionics Maintenance Conference steering committee since 2004, serving as chairman for the past four years. The committee is part of ARINC Industry Activities, which is currently part of the Association of Automotive Engineers (SAE International). In addition, I am a technical director of the AMSTEL-M company, which develops aircraft modifications for Russian flight operators. Awards: VOLARE Award, 2004, TQP Award, 2004, Nominations: Aerospace Journalist of the Year Award, 2004 and 2005

Looking to the future, what are the key challenges you see in the aviation sector with regards to obsolescence management?

Airplanes are built to last! Generally speaking, an aircraft today is expected to fly 20 to 30 years. Twenty years ago, airplanes were designed for an even longer life, and all materials and equipment were built to minimise obsolescence during the aircraft's lifetime. Modern airplanes are built to be more cost-effective. Modern technologies are now embedded in the aircraft, and although there are contracts that require obsolescence management, problems arise when companies go bankrupt or when sub-suppliers are acquired by competitors, change their names, and stop the production of key components. Modern commercial technologies, however, will not last 20 or more years. A realistic life cycle is, most of the time, around seven years (though there are exceptions). Besides equipment, software also plays an important role. Even if there is software available in software libraries, it is not possible to program it into integrated circuits (EPROMs) because they are no longer available on the market. Even if there are compatible chips, they must be qualified for use in aircraft equipment. This makes manufacturing costly and time-consuming.

Paint a "perfect world" scenario for meeting these challenges. What would it look like?

In the first place, we need purchasing contracts which will protect buyers from obsolescence problems during the life cycle of each component. Standard contracts hardly regulate anything that will happen 20 years from now. Also, each supplier of components must have an obsolescence management plan. Some aircraft manufacturers are now paying more attention and are demanding obsolescence planning from their subsuppliers. The next step should be that if a supplier declares its product to be obsolete, it should then immediately make the detail specifications of the product available to enable somebody else to step into the business and manufacture the parts. That should be regulated with legislation.

How realistic do you think it is? What needs to happen for this to occur?

We need a lot of cooperation to solve these problems. For example, 286 old computers are worth nothing to one company but a fortune for another. It's all about cash. Legislation might help. Unfortunately, some companies purposefully introduce obsolescence just to be able to sell new equipment which is not compatible with the old equipment. The purpose of that is to increase profits. We see more obsolescence in a 787 aircraft which has been on the market for just three years than in a 747-400 which has been around for more than 20 years. In addition, in modern aircraft, airframes are using too much commercial offthe- shelf components and technologies.

FOM Amsterdam 2016 will be sharing best practices from different high-reliability verticals – from medical to aerospace and defence, transport, etc. What have you learned from other business sectors in regards to obsolescence management? Please provide an example.

Prevent the use of commercial equipment in the aircraft, and manage software very accurately. Make sure that you anticipate market developments. If you need piece parts, calculate how much you will need to support your equipment for at least the next five years. I have learned that people can become obsolete too. In my area we are trying to document procedures as much as possible. We also train people to understand job functions and be able to take them over if somebody leaves or retires.

What do you think the aviation sector can bring to other markets in terms of innovation?

Regarding obsolescence, aviation can provide solid examples of regulations, procedures, obsolescence management, and guidance materials which can be used elsewhere. Normal procedure in aviation is to announce the obsolescence two years in advance and provide users a fair chance to engage in a last-time buy. Specifications and procedures for how to qualify materials should also be well defined. It is time-consuming, but if procedures are defined, that is better than having no procedures at all.

Can you briefly tell our readers about the AMC 2016 event in Atlanta and your involvement in it?

The 67th AMC conference will take place in Atlanta from 25 to 28 April 2016. This time it is hosted by Delta Airlines. People from all over the world will join in the meeting room, excited to solve problems, earn cash, save money, and learn many new things. This is how the AMC conference works: Step one is to submit questions in advance. Everyone is asked to submit questions, but there is one simple rule: discuss your problems with air-framers and suppliers all year long beforehand. If they are not cooperative in solving those problems, be sure to submit those questions. Each year we collect approximately 200 to 250 questions. The questions are published in February, and there is a window of time for suppliers to prepare their answers. While somebody is working to answer your questions, you also have the opportunity to ead others' questions. You might learn new things, find that you have a similar problem, or discover that you have a solution for a problem submitted by another. All the questions and answers will be discussed during our conference. At the end of each day, there will be a seminar including topics that the engineers have selected for the conference. The quality of topics and presenters is amazing. It is an honour and a noble task to share this knowledge with others. Besides the official part of the conference, there are some interesting collateral items. There is the AAI reception on the evening of Tuesday 26 April. This is a so-called must attend event. For some it is the most important part of the conference. Suppliers show off their new products, such as new DER repairs, new PMA parts, or solutions for obsolescence, for example. This is the place to be if you want to learn how to save money for your company. It is the "hottest" evening event of the conference. The conference is organised by a small group of people, the AMC steering group. This group manages the event and activities, including the hotel and the 700 conference attendees, and makes sure everything runs like a well-oiled machine. Also, these same individuals make sure that the conference stays within budget and delivers attendees the greatest experience they can have, now for the 67th time. I am more than honoured and privileged to be a chairman of this group.

Going back to a question a member of the audience asked our panel at FOM 2015, if you had one wish for the future of obsolescence management, what would it be?

My wish would be to create awareness, because obsolescence is here to stay whether we like it or not. Obsolescence will never go away. Therefore, we should learn to anticipate it! It is easier to be prepared than to let obsolescence overcome you. Anticipation is the first step in obsolescence management.

Welcome note in AMC program for 2016!

On behalf of the AMC Steering Group, we look forward to seeing you for the AMC | AEEC in Atlanta, April 2016, hosted by Delta Air Lines. This will be the 67th annual meeting of this conference.

The only constant factor in aviation industry is change. The period since the last AMC was no exception. Many B787s and A380s are flying. A350 is entering the theater too. Life will never be the same because new birds are different. We are already used to idea that MD-11 and B747-400 are leaving or going to leave the skies. We are looking forward to greet the A320-NEO and B777X in the near future.

Airframers and OEMs are doing a great job to deliver reliable and safe aircraft, which will spend the majority of useful life in hands of operators. Actually they delivered a cash generator to airlines but it is not that easy.

The 67th Annual AMC is collocated with the AEEC General Session.

The difficult part is to make sure that airlines effectively use those money generators. In another words, we are at the turning point. New aircraft, new components, new technologies, new contracts, new playing field. Everybody must change. The most terrible move you can do is to stick in the old pattern and do nothing, thinking that the wind of change will blow over. Unfortunately, the wind of change keeps blowing and nobody can stop. Therefore, respect new rules, business cases, and environment and adapt. The ones who are not able or not willing to adapt will vanish just like dinosaurs vanished 75 million years ago. They will go and will be forgotten.

One factor which can point the way is the incredible brain power which can only be experienced at AMC conference. This is not just a conference: it is THE Conference. You can't find a better place and you can't, definitely can't, find a place with more knowledge about modern aviation.

The professionals of SAE-ITC improved Industry Activities and made it stronger, which is good! Airlines and OEMs are there to elaborate, discuss problems, provide solutions, and give unbiased advice. The only thing you should do is to come over to Atlanta and explain what is bothering you. Between 650 professionals, there is always at least one who can help you.

Even if you didn't submit your question in advance, just go to an airframer representative, supplier, or airline engineer and ask. People will help you because they care. They want to help. They are willing and they are able to help. Just ask. Today they will help you, tomorrow you might help them. That is the spirit of the conference.

Welcome to AMC 2016!

The AMC brings 600+ industry professionals together to discuss avionics engineering and maintenance.

At AMC, we also care that everybody is educated; therefore, we organize seminars. You could say that seminars are customized just for you, me, and a broad audience. It is not a coincidence that we issue a questionnaire at every conference. Based on those inputs, we carefully select seminars which are of most interest. This time there is no exception. On your request, we attracted the best speakers in the industry to provide unbiased, honest, and high quality information just for you. The information is worth money and if you use it intelligently, it can generate a lot of cash for your company.

We sincerely appreciate the continued support from our Member Organizations and Corporate Sponsors of Industry Activities that enable us to conduct this one of a kind global event for the benefit of all our attendees. It is our duty as operators, OEMs, and airframe manufacturers to spread the word about who we are and what we do. As a supplier or airframer, please remember that being an AAI member is an important contribution to AMC, but if your organization has experienced increasing success as a result of participating in AMC, please consider also becoming a Corporate Sponsor of Industry Activities. By growing our memberships and sponsorships, we can ensure the future of AMC as your primary source for constructively resolving technical issues, providing information exchange, and networking with business partners you might only see this time of year.

To suggest a symposium topic, contact a member of the AMC Steering Group.

OEMs, of course, have their justification for attending the AMC conference. It is a unique opportunity to meet and greet more than 45 airlines. It is a dramatic cost saving to go to AMC instead of traveling to 45 or more destinations and spend many days on a trip. If they plan it well, they can meet 45 airlines in just 4 days.

Airlines can meet 150 OEMs in the same period of time instead of having 150 meetings spread across typically a six month period. Actually, everybody wins. Which is excellent.

Also, I would like to express my great respect and appreciation for Delta airlines which is hosting the conference. Delta Air Lines' engineers know the benefits and know the challenges of the aviation industry and besides all hard work to make their airline one of the best and most respected on the planet, they also spend their additional time and effort to host the conference. I would like to say big thank you for all good care: THANK YOU DELTA AIR LINES.

Finally, I would like to take this opportunity to thank the ARINC IA staff and my steering group associates for the additional effort to make the AMC happen for the 67th consecutive year. My steering committee members are delivering additional effort on the top of their normal airline activities to help organize and improve the conference. Their companies are convinced that there is a good payoff for all the effort because they realize that the more you put into ARINC IA, the more you benefit. Thank you all for being so cooperative.

Saying that, I would like to remind you that Atlanta is the place and April 24, 2016, is the time.

I will be there and make sure that you will be there too.

Atlanta will be the hottest place in the universe, make sure that you are there too.

AMC Opening Speech ATLANTA 2016

Good morning my fellow engineers!

My name is Marijan Jozic, and I am the chairman of the AMC Conference. I hope that you all have found some time to see the beautiful city of Atlanta. On behalf of AMC and AEEC Steering, welcome to the 67th AMC | AEEC Conference. Welcome to Atlanta! Thank you, Delta Air Lines, for being the host.

You know, running a maintenance department at an airline is one immense and frustrating process. I often ask myself, why am I in this business? Why are we in this business? Why would anyone in his or her right mind get into the aviation business?

To some, it is a job. To others, it is the thrill of watching the aircraft fly away, knowing you personally played a part in it. To many of us, more than in any other industry, it is the passion because you are an aviation romantic who still thinks it is an incredible experience to see any aircraft leaving the ground. In more than 50 years, the business has seen a huge collective net loss. The difference between profit and loss is razor thin. The cost of the man hours of highly educated technicians with a lot of experience is lower than the cost of the man hours of a plumber or a painter. The responsibilities, of course, are much bigger. It is obvious that MROs can hardly make money that way. If you look at the cost of material and man hours, the accountant might figure out that perhaps a modification can generate some tiny profit. But there are many hidden costs. Think of all the time spent in meetings, telephone calls, with contracts, customers, purchasing, Service Bulletin planning, engineering, production units, planning departments, preparation departments, quality assurance, DRs, accounting, calculators, flight ops, vendors, incoming goods, inspectors, and suppliers. Then we have preliminary design reviews, critical design reviews, technical coordination meetings, first article inspection, and WebEx conferences. For all those hours, if they are added to the project cost, it is obvious that profit is infinitesimally small.

So why are we staying in this business? How can you explain it?
Finally, somebody pointed out to me what it is. He pointed out what we all have in common:
It is more than obvious that we all have severe brain damage.
Yes, you heard me right. Severe brain damage. Otherwise, how can you explain this?
The business is:

- Very capital intensive: expensive parts and ground time.
- Fuel intensive: prices widely fluctuate in a blink of an eye.
- Labor intensive: work until you drop – the aircraft must leave as soon as possible.
- Customer service intensive: otherwise, pay a penalty.
- Intensely operational: late delivery means cancelation of the flight, high penalties for delays and cancelations.
- Perishable inventory: even worse than bananas.
- Competitive: man hour price under great pressure.
- Highly accurate: no mistakes allowed.
- Obsolescence sensitive.
- Highly regulated, with millions of procedures.
  Highly taxt.
  - Extremely sensitive to events like SARS, Ebola, Zika virus, and of course the 9/11 and Brussels attacks.

- Under extreme supervision over advertence authorities: think of all the audits.
- Intellectual property is protected and legally controlled.

The good news is that our severe brain damage is not lethal. But yes, it is true: when everything is finished, and an aircraft leaves in perfect condition, it is an extreme thrill and we all love it.

So let's hammer the start of the 67th AMC!

Atlanta 2016 Closing Remarks

AMC Chairman

I would like to thank you all for being here and solving problems and working very hard on questions. We have had an exciting 4 days and nights, and have really accomplished a lot of good things. A few observations and statistics:

- We had 725 attendees at the conference.
- Accounting the last Honeywell success story, it will be a total of 19 success stories.
- We are coming from 25 different countries all over the globe.
- We have encountered one drag race, because they were recording a movie.
- We have 36 open items that will be handled at the next conference.
- We had a visit of FAA, a presentation of FAA that is the first time in the last 20 years.
- We had 7 microphone resets.
- We had 5 camera operators and had 4 annoying lights at the back. You do not see them, but we were suffering here on the stage.
- We have an increase of female engineers by 100%, with respect to the last meeting, which is good.
- Last but not least: We have 100% compliance with ARINC IA advice in the program. ARINC IA was advising not to walk from the airport to the hotel. Nobody was walking from the airport to the hotel, so that is 100% compliance.

Finally, thank you to AAI for excellent food and taking good care of us in the conference. Also, thank you to the OEMs and airframers for their hard work and contribution in answering our questions. And congrats to the awarded individuals.

Have a safe trip home, and see you in Milwaukee in 2017.

Now, we can hammer the closing of the conference.

Feel the Burn!

On April 21, 2016, just before I departed to Atlanta to the 67th AMC conference, I made a note on my Facebook page:

"I am going to the hottest place in the universe." I didn't mention Atlanta or the AMC. I got a few likes and a couple comments: "Hey man! Are you going to AMC"?

Some people are already used to the idea that each year in April or May, there will be an exceptionally hot event called the AMC. There is no better place for an avionics engineer than AMC. This year was no exception. Engineers from all over the globe (25 countries!) traveled to Atlanta. The AMC is a lot of work, but by this time most of it is already done. Engineers have already submitted their questions back in January. After that everybody started to do their own preparation. Some of the engineers were answering the questions and some were preparing for the event. More than a thousand greatest avionics brains were getting hotter and hotter. That all culminated on the 25th of April 2016.

725 people came to the spot. This time the spot was the giant ballroom at the Hyatt Regency Hotel in Atlanta. First you could feel the heat, and then the burn. For four days we collaborated, talked, and invented new concepts, products, and services.

Nearly two floors below the street level, we almost experienced a melt down. Everybody was energized and let me explain why.

It started with excellent keynote speeches and well-deserved awards for some outstanding engineers. It is a great tradition that we remain absolutely silent about award winners. The individuals who received the AAI Volare, AEEC Trumbull, and AMC Roger S. Goldberg awards were chosen by their peer engineers weeks in advance. The secret was kept till the last moment to surprise them. That was a once in a life time experience for each of them. Nobody expected it. Some of the recipients thought that they didn't deserve it. This humility is the typical engineer's attitude. Engineers are constantly striving for perfection and nothing is ever good enough. They always see room for improvement, but they are also underestimating their own ability. They think that they can do a better job even when its at the limits of performance. The awards ceremony is always a great show. Feel the burn!

The AMC open forum discussion was great as always. There were 240 questions resolved or coordinated and we had nearly 20 true success stories which is an excellent score. Some questions remained open because there was not enough time to resolve it. Without a doubt they will be fixed later. Likewise, many of the open items from last year's AMC were closed showing our strength in the industry. Of course that we had a great dialogue while discussing the questions. The great tradition of the AMC continued. Boeing and Airbus, Honeywell and Rockwell Collins, UTAS and Thales, they were all helping each other to reach the ultimate target to solve difficult engineering problems and make the aviation industry a better place. Those who were not present, can always profit from our work. The AMC attendees have made the correct decision for them. Some not in attendance may not like it. But they were not in Atlanta and did not have any vote in making the decisions. Now they can only follow. I much more like to lead rather than follow. Therefore, I go to each and every AMC. I want to feel the burn!

And then there were the presentations. This year, as always, we had the superb quality of presentations during our symposiums. Our first seminar was about aircraft health management. Do you remember the movie: Space Odyssey 2001? Do you remember the computer HAL 9000? At certain moment HAL reported to Bowman the imminent failure of an antenna control device. Of course, Arthur Clark have made the big story of it 50 years ago but the idea is that computer will predict the failure of a device in the spacecraft. The future is now. In a majority of cases, we can predict failure and allow the replacement of the component before it even starts to malfunction. This means that no revenue time is lost. Is that not wonderful? If we manage to predict failures and act timely, we can prevent many of our operational problems. Great ideas! Feel the burn!

On Tuesday, we had presentations about PMA parts. I don't recall that we have ever had the FAA providing a presentation at the seminar. If we did, it was a long, long time ago. We had great presentations from Robert Sprayberry with the FAA, Michael Rennick from Delta (a 2016 AAI Volare award winner), and Patrick Markham from Heico. In 2015, we had PMA presentations in Prague and our AMC audience asked for more. Well, we managed to provide more PMA discussion without repeating the same content from last year. There was a lot of new information and a lot of really useful stuff. It seems that at our AMCs we have also started to educate. This is very important item not to forget. Our three missions are: solving problems in the open forum, adopt and implement new standards and guidelines, and education. People attending our conference are getting a full package. There is no other conference which can even come close to it. Feel the burn!!

Finally, just to mention that there is a desire to redefine avionics. In new environment when A380, A350, B787, and B747-8 aircraft are flying we should not close our eyes and say avionics that is traditionally defined by ATA 23, 24, 31 and 34, etc. because it is not. Nowadays you can hardly find an ATA chapter that does not have electricity. Every regulator valve, check valve, actuator, servo, or air duct has sensors and/or microcontrollers installed. The new definition of avionics can be everything with a wire attached to it. Well in that case, avionics is taking over the whole airplane. The only part without a wire is perhaps the fuselage. Honestly what is left over without wires? Not much.

In the end, I hope that all 725 people who attended the AMC had great time and safe trip back home. Atlanta will forever stay in our memory as the year that we Felt the Burn!

Make Avionics Great Again!

If you have ever been in a room full of engineers, you know giving them a problem to solve is like throwing a meaty bone to a dog. This is what they thrive for. The culture of relentless innovation is driven by airlines, avionics OEMs, and airframers. If you look back at the last 20 years of aviation you can see that we are hitting double-digit efficiency gains on each new model of aircraft. Airframers from all corners of the globe (Airbus, Boeing, Embraer, and Bombardier) are doing a great job together with avionics OEMs. This is just the beginning of the life cycle of new birds. The new models are handed over to the airlines engineers who will take care of that aircraft for life, operating for another 20 or 30 years. Everything is one big, smooth running machine. Airframe and avionics reliability is extremely high. Fuel consumption is low. Aircraft can fly much longer than ever before. If there is a problem, simply send your engineers to the AMC and just throw them a bone. This year was no exception. We had 18 great success stories and a lot of smaller success stories. One of those stories is about the ULB (Underwater Locator Beacon). During the open forum discussion, we had an intense discussion. As we learned, discussions about the ULB can take quite a bit of time.

Basically, the issue is that there is a new 90 day ULB which will replace the 30 day ULB. At face value, this is a success as the technical solution has been designed to allow a longer inspection time. Now to work out the details about part numbers, certification, CMM, etc. At the AMC, the forum decided to have an engineer's side meeting. The proverbial meaty bone was thrown to these engineers. Honeywell graciously offered a meeting room and we set the meeting time.

And here we are. A room full of engineers fighting to solve the ULB issue. When I saw that scene I got tears of happiness in my eyes. That is the true spirit of AMC. This is how we make avionics great again. The meeting was not long – 20 minutes' maximum. And we made a solid plan. That is how the AMC works. Look at the picture. This is the group of engineers who were fixing the problem for the whole planet. We decided what to do and how fix the part numbers and the difficult configuration management issue of ULBs on every single aircraft flying. If you were not there, too bad. You might not like our solution but it is too late. Next time you can be part of it if you become a ARINC IA member and get access to the AMC networking, industry standards, and this great bunch of people. In my opinion, these are the most willing and helpful engineers on the planet.

An additional remark: in the last couple of years we have been in a learning mode. We have encountered the introduction of new business models, new aircraft, and the lawyers' penetration into our environment. And at certain moments it was uncomfortable. But we overcame the obstacles and now it is time to make avionics great again. Talk to your peers in industry, engineers at another companies, etc. Let's spread the word and invite them to become an ARINC IA supporter. There are a lot of brilliant minds.

We need them and we will be more than happy to get them on board. Make Avionics great again, make history and help us to lead the industry. Believe me it is better to lead than to follow. If you lead you will have that great feeling. If you follow, it is depressing. That is why in the AMC we lead!

All my Bags are Packed - I'm Ready to Go

2017 AMC Program

Avionics Maintenance Conference

May 1-4
Milwaukee, Wisconsin USA

I am looking forward to see you at the AEEC|AMC in Milwaukee, Wisconsin in May 2017, hosted by Carlisle Interconnect Technologies. This will be the 68th annual meeting of this conference.

In my house cleaning speech in Atlanta I will tell the young engineers who are, attending the AMC conference for the first time: "Once you are here you can't go back. The AEEC|AMC conference is an addiction. Not because it is in different city every time, but because it is extremely useful".

My addiction started at the 50th AMC Conference in Baltimore. Since then, I have been a big fan of the conference. It is really very rewarding. Therefore, immediately after closing the current year's

AMC conference, I start the preparations for the next one. It is not just packing my bags. That is the easy part. In my opinion, one particular part of packing is much, much more important.

Since this will be my 18th AMC that I have attended, I should give you some advice. This is my experience talking, and I will start by expressing it in an anecdote. Many years ago a Croatian singer Arsen Dedic said in the interview: "We are learning from our mistakes, but the British are learning in Oxford".

I found it a smart statement and decided from that time on to listen to experienced people and their advice. It is far better than to learn from only your own personal mistakes. (Well, not always but now and then.). I don't really have to follow their advice but just to consider their missteps and that I might learn something and gain some advantage.

So here is the top 2:

1. You can never plan on every eventuality, so don't pack for it!

2. Burkina Faso also has shops and so does Milwaukee.

Basically you need: Luggage, Clothing, Toiletries, and Technology. It is not more than that and there is no use to spend more time in this article about that.

And now the important stuff. Going to the next AMC is a journey which starts two hours after you return from the just past AMC. In the days before the AMC report is published make sure that you notified your organization of your own success stories. That is a very important point. It will give you justification for making the trip to your next AMC.

You met a lot of people, that is for certain. But, what did you learn? Those learning moments are important and they are also bringing cash to your organization and boosting your own confidence as well.

I will give you an example. Many years ago were considering to invest in an ATEC6 series Automatic Test Equipment (ATE). Without much experience with this type of purchase we saw a lot of issues along the way. One of the worries was how to get the electrical power to the machine. Would we need larger transformers or special powerlines? How much power does an ATEC6 consume? How do we transport ATEC6 and all associated equipment through our building? Do we need special flooring reinforcement? And so on. Then, at AMC I saw that they had a full blown ATEC6 series in hotel room on the 4th floor and they connected to power through the ordinary power socket in the hotel room.

In just one blink of my engineering eye, all my worries vanished. If those guys managed to install ATEC6 in a hotel room and it was working, we certainly will manage to install it in our facility. No big deal. That saved me many meetings, and the worries convincing my management that it is doable.

Knowing our capabilities gave me full confidence. Yes you are right, this is a trivial example but this is how it sometimes works at the AMC. Just one tiny little thing and your comfort level is high and you become unbeatable because you possess the information.

Back to my storyline. After the AMC Report is published make sure that you provide it to the people in your organization who were asking questions. They will become your best friend if you fixed their problems. They do not primarily worry about the money savings. They mostly care about a resolution to their problems. The more problems you resolve the more popular you become. That will give you a travel visa to the next AMC.

The next step is to open a new folder in your computer and drop in questions for next AMC. Be patient and create a document for every question. In a couple of months you will have many documents in the folder. Some of those questions might be solved before February 2017. Just delete those and keep the unsolved ones. Those are the questions which must be submitted on behalf of your company.

One more side story about questions: Many years ago I attended a Satcom course in Phoenix, Arizona. And did you know who is in Phoenix? Our friends from Honeywell. I arrived on Saturday and decided to go to the Kitt Peak Observatory because my course started on Monday. That is without a doubt a beautiful place. There was a tour guide who said: "Do not hesitate to ask questions. Ask about anything. There are no stupid questions. The only stupid question is the one that is never asked."

The point is to ask a question even though it might look trivial. If it is bothering you just ask. Think of Carl Sagan who once said: "There are naive questions, tedious questions, ill-phrased questions, questions put after inadequate self-criticism. But every question is a cry to understand the world. There is no such thing as a dumb question".

And last but not least: sending suggestion emails to the AMC Steering Group if you want any particular aviation topic discussed or presented at the AMC symposiums. Those emails should be submitted before October 2016. The AMC Steering Group will then review all submissions and decide which topics will be presented. We are very successful in getting the best speakers in the industry. Your idea could be chosen so we all can benefit. Also, if you wish to present about any topic, make it known to AMC Steering Group . We on the AMC Steering Group have much more to do before the next AMC but that is a completely different story. The fact is that Milwaukee will be the place and May 2017 will be the time. Do not miss it. Roger Goldberg said, "It is your conference, it is what you make of it."

We need you and you need us. Together we are strong and together we can improve the aviation industry and make it safer. See you In Milwaukee.

What to inspect at the 2017 AMC / AEEC conference in Milwaukee?

(Interview with Avionics Magazine beofore Milwaukee conference)

Three new ARINC avionics standards and eight supplements to existing standards have reached an industry consensus for the 2017 annual Airlines Electronic Engineering Committee (AEEC) and Avionics Maintenance Committee (AMC) general session. Also to be discussed are Boeing 787 HF receiver issues, the European VHF Data Link Mode 2 (VDL Mode 2) infrastructure and global aircraft tracking.

Ahead of the conference, Avionics caught up with Paul Prisaznuk, executive secretary and program director for the AEEC, and Marijan Jozic, a European airline avionics maintenance engineer and the chairman of AMC. Both provided updates on key communications, navigation and surveillance (CNS) as well as air traffic management issues that will be highlighted during the annual conference.

AEEC and AMC are organized to function simultaneously in separate sessions, with AEEC noting its mission as developing engineering standards for "avionics, networks, and cabin systems that foster increased efficiency and reduced life cycle costs." AMC notes its objectives as promoting "reliability" and reducing air transport avionics operational life cycle costs by "improving maintenance and support techniques through the exchange of technical information."

Among the top avionics issues up for ARINC standardization at AEEC 2017 are the VDL Mode 2 infrastructure issues for the European air-to-ground data communications protocol. After the ELSA program provided its report on the issues that lead to a delay in the 2015 mandate, the AEEC executive committee's agenda includes two items specific to VDL mode 2.

"First, a mature Supplement 7 to ARINC Specification 631: VHF Digital Link (VDL) Mode 2 Implementation Provisions, includes VDL-2 multi-frequency management and changes as a result to the European data link services implementation rule," said Prisaznuk. "Looking further down the road, the AEEC Executive Committee will consider adding a Connectionless VDL-2 Protocol Variant to ARINC Specification 631."

The connectionless protocol will create efficiencies in the link set-up time and improve VHF channel utilization. The work package to prepare the connectionless VDL-2 protocol is expected to be completed in 2019, he added.

AeroMACS, a new broadband data link that has the ability to support the ever-expanding range of air traffic management communications technologies emerging under the modernization initiatives of NextGen and Single European Sky, is already installed in several airports in the U.S. During the AEEC general session, ARINC Project Paper 766, which focuses on AeroMACS Transceiver and Aircraft Installation Standards, will be up for adoption. Prisaznuk notes this is a leap forward in ground-ground communication on the airport surface. AeroMACS is based on the IEEE 802.16 family of WiMAX protocols and can achieve 3Mbps data throughput (compared to 31kbps for VDL-2).

"AEEC's action in Milwaukee will send a clear signal to the world's airports that the airlines are prepared to add AeroMACS radios to their aircraft once the ground infrastructure is in place and business case makes sense. The AEEC Executive Committee will consider a new activity to introduce IPv6 to aircraft and avionics," said Prisaznuk.

IPv6 is expected to apply to safety services and non-safety services alike. An IPv6 roadmap is expected to be available in 2019, he said.

A security overlay for the Inmarsat SwiftBroadband Safety system (SB-S) is also now mature and ready for AEEC approval, according to the AEEC executive committee member. Additionally, a fifth supplement to ARINC 622, which standardizes air traffic services data link applications over the ACARS air-to-ground network is also ready for approval from the executive committee. AEEC is working with FAA on this activity.

Finally, among the new and emerging topics to be discussed in the general session will include unmanned aircraft systems technologies. General Atomics will provide a presentation at the session on this topic.

The annual AMC session always features an open forum style, where airline maintenance engineers discuss current challenges and issues they're encountering regularly with in-service commercial aircraft avionics hardware and software. Jozic said he is looking forward to a discussion on the contamination of Boeing 787 HF receivers by a leaking water issue within the aircraft that many airlines maintenance engineers are experiencing right now.

"The water is leaking on HF receivers and destroying power supply," said Jozic. "Every burned power supply brings replacement costs of $15,000 to operators."

Two new standards have also been adopted by the AMC committee within the last year, including the ARINC 422 guidance for avionics service bulletins and modification status as well as the guidance for assignment, accomplishment and reporting of engineering investigation for aircraft components. There will also be new standards adopted by the AMC steering committee in a meeting before the conference, including aircraft support data management ARINC 675 and field loadable software ARINC 667.

AMC also plans on launching the mechanical maintenance conference later this year, giving mechanical engineers a chance to meet their peers and resolve issues in the same way AMC does with its annual meeting.

Milwaukee at 2017 AMC!

Note from the chairman for 2017 AMC Program

On behalf of the AMC Steering Group, we look forward to seeing you for the AMC | AEEC in Milwaukee, May 1-4, 2017, hosted by Carlisle. This will be the 68th annual meeting of this conference.

During the Christmas Season, I visited New York, better known as Big Apple. In front of Trump Tower, I almost bumped into George Clooney. In my confusion and surprise, I asked George: "What is the highlight of 2017?" I expected that he would say, "Nespresso, what else!" But he said to me – "AMC, what else!"

I agreed! Let me explain why I agreed. The AMC conference is not just a conference. It is a movement. We call it the AMC conference but we do more. AMC is an acronym for Avionics Maintenance Conference. Outsiders would say that we are only doing a conference, but that is just part of the truth. Besides the conference, we develop standards. Many of the ARINC standards, not all but many of them, are designed and issued by engineers connected to AMC the conference. That is the second pylon of the AMC movement. The third pylon is education. Yes, we also educate. Engineers who follow the AMC movement will agree that we from AMC contributed to the creation of the knowledge base for several big subjects in modern aviation. Let me just touch the base with some of them.

Future Obsolescence Management is the area which used to be just something else. We created awareness and lifted the know-how of engineers to be able to manage obsolescence. In the good old days, we just accepted the obsolescence and were sorry for ourselves that we were having problems. Now we know that obsolescence can be managed. We also know that airframers like Airbus and Boeing have obsolescence management plans and that they interrogate their supplier twice a year about possible obsolescence of their products. We know that we must incorporate an obsolescence management plan in each of our contracts. Also, we know that there are fifty shades of gray but only seven shades of obsolescence. Obsolescence of: LRUs, piece parts, test equipment, chemicals, documentation, software, and people.

We also have a Rule-of-Three. For engineers among us who are not familiar with term Rule-of-Three, here is the explanation: the Rule-of-Three is a writing principle that suggests that things that come in threes are funnier, more satisfying, or more effective than other numbers of things. The reader or audience of this form of text is also thereby more likely to remember the information. This is

because having three entities combines both brevity and rhythm with having the smallest amount of information to create a pattern. I will give you some examples:

- Sex, drugs, rock and roll (we know this one very well)
- Friends, Romans, Countryman (William Shakespeare)
- Stop, Look, Listen (Public Safety)
- Faster, Higher, Stronger (Olympic Games)
- The good, the bad, the ugly! (in the film industry)
- Sex, lies, videotapes (in the film industry)
- Lies, damned lies, and statistics. (There are three kinds of lies)
- Ho, Ho, Ho (this one is for Christmas)
- And the last one, which I heard from Ray Frelk:
  There are three kinds of people: those who make things happen, those who watch things happen, and those who wander what the hell is going on.

Enough examples. The rule of three for obsolescence in aviation is: prevent, detect, solve. You might call it the sex, drugs, and rock and roll of obsolescence: prevent, detect, solve! Knowing that when we design a new aircraft and system, we should prevent obsolescence. The whole lifetime of an aircraft, we should monitor the situation (that is what Boeing and Airbus are doing twice a year) and if obsolescence is predicted, we should solve it. Prevent, detect, solve!

Review ARINC Report 662 (obsolescence management) and be aware that in 2017, there will be some activity in that area to make it even better.

The next example of how we educate our engineers is contract management. Not so long ago, we had an easy life. The lawyers didn't bother and interfere with engineering activities. Suddenly, we were in the middle of an intellectual properties war. OEMs changed the policies and we, operators, were very confused. At certain moments, there was tension and we almost become enemies. Emotions were heating up and we even heard famous words: "If I am going to jail, I will take you with me!"

Yesterday, business partners were not business partners any more. The tension was high, but we from the AMC movement managed to learn new things. We organized many seminars and discussions to try to understand each other. Eventually, we developed a standard called SCEA – Standard for Cost Effective Acquisition. Through open discussion at the conference, we created awareness that there are Product Support Agreements which protect us from many unwanted issues. Many of us were not even aware that Boeing and Airbus had Product Support Agreements. Suddenly, if you do it right, a new world is in front of you. The PSA is for SFE (Seller Furnished Equipment) and SCEA is to be used for contracts with suppliers of BFE (Buyer Furnished Equipment). SCEA is very important document, known as ARINC 674.

Next working area is PMA (Parts Manufacturer Approval). We were very successful in organizing seminars and bringing the best experts in the world together to speak at the AMC conference. Two years in a row, we had hundreds of people in the meeting room listening to speeches from FAA, HEICO, Delta Air Lines, and Barfield. All of them, leading experts, were there to share their knowledge and to educate.

Just check last year's Plane Talk and read the transcript of discussions after the presentations. I have the feeling that we could have stayed in the meeting room for another hour or two until our questions and discussions were exhausted.

Finally, just to mention some topics which are probably coming back in coming years, we already had a seminar about 3D printing. It is safe to say that AMC is for sure the first big aviation conference which handles 3D printing. We provided the base for new discussions and ideas. Soon, the seminar about chemicals will be on our agenda. Also, not to forget predictive maintenance, which is becoming a fact of life and is closely connected to big data. We are at the edge of a new era. Just like every new era, it started with chaos and little by little, through AMC movement, we are bringing order. Some people are already in panic saying we can't handle this complicated new type of aircraft. For all of them, I have a massage: Yes, we can! We will make avionics great again.

And now back to George Clooney. I was really in New York, and I was really in Trump Tower, but I didn't bump into Clooney. And my wife didn't bump into George either (although she likes his hairstyle better than mine). That story about Clooney was fake news. All other things are real.

I just wanted to get your attention and tell you wonderful things about AMC movement.

Welcome to the 2017 MMC!

Written for Inagural MMC (Mechanical Maintenance Commitee) conference in Clevelend

My name is Marijan Jozic. I have been the AMC Chairman for the past six years, and now have the pleasure to be the first chairman of the Mechanical Maintenance Conference (MMC). I have been involved in ARINC Industry Activities (ARINC IA) over the last 20 years.

For the first time in the history of aviation, we will meet at the Mechanical Maintenance Conference. The MMC has been a topic of discussion within the avionics community for some time. Mechanical engineers were asking: "Why? Why don't we have an event like the Avionics Maintenance Conference (AMC) for mechanical systems?"

It is a magnificent formula. Airline engineers from all over the world submit their questions to ARINC Industry Activities. The questions are shared with OEMs and airframers and are issued in the program for the conference. Everybody involved can study the questions and prepare answers. At the conference, we handle all those questions and count success stories.

The AMC Steering Group used this formula in the creation of a new mechanical conference: the MMC. The MMC is an air transport industry activity serving the promotion of mechanical systems, equipment reliability, and performance. It is the medium for the exchange of information among users, repair facilities, installers, suppliers, manufacturers, and designers of avionics systems and components.

But there is more to the MMC besides the open forum, where we answer the submitted questions. Just like the AMC, this activity is supported by additional pylons. One such pylon is education through seminars.

Last year, the AMC held a seminar about Parts Manufacturer Approval (PMA). We recruited the best experts in the industry to talk about PMA parts. To maintain objectivity, we looked at the topic from different angles. Legal people are familiar with the Latin phrase, audiatur et altera pars, which means, let's hear the other side! To comply with this concept, presenters represented the FAA (regulatory authority), Delta Air Lines (operator), and HEICO (manufacturer). Of course, everybody is free to ask questions during the seminar and discuss every bit of the PMA subject. The seminar was such a success that we are going to repeat it for the MMC engineers at the first MMC.

The final pylon of ARINC Industry Activities is the design of technical standards. Yes, we produce standards, too! ARINC Industry Activities has 544 standards in circulation. It is not because it is easy, but because it is hard and it is necessary.

Let me explain! ARINC Industry Activities has three activities (and now, with the introduction of the MMC, four). These activities are:

- AEEC (Airlines Electronic Engineering Committee)
- AMC (Avionics Maintenance Conference)
- FSEMC (Flight Simulator Engineering and Maintenance Conference)
- MMC (Mechanical Maintenance Conference)

Each of these activities organizes working groups, which come together to design standards. Together, we update or issue about 40 standards each year. These standards are necessary, and it is hard work to create and maintain them.

Airlines who are not a member of AMC, AEEC, FSEMC, or MMC must either accept our standards or reinvent the wheel for themselves. That means that they are not leading like we are, but following. Believe me, it is better to lead than to follow.

That is what we do: we lead and point the way. We want you to go home from the MMC with a satisfactory solution to your problem, a lot of energy, and new knowledge. If you are not satisfied with answers given at the open forum, just make it loud and clear that you want to leave the item open and OEMs can work on delivering a satisfactory answer after the conference. We want you to report back to ARINC IA when the problem has been solved so we can close the question before the next conference.

Other things, like obtaining new knowledge and meeting OEMs and airlines, are in your hands. I have only one piece of guidance about that: The more you put into it, the more you gain. Make a short list of subjects you want to discuss with peer engineers or OEMs. Plan your work at the MMC, and work your plan. That is how you will win.

At the next conference (I am sure that you will come back!) you will tell me, "Mr. Chairman! I am winning so much at the MMC that I can't take it anymore." I will tell you, "No, we are not stopping! We have to win more to make our industry safe, cost effective, and extremely reliable."

See you in Cleveland!

Opening Remarks for Clevelend MMC 2017

Ladies and gentlemen, my fellow engineers. I hope that you have found some time to see the beautiful city of Cleveland. On behalf of the AMC Steering Committee, I welcome you to the very first MMC Conference. Welcome to Cleveland!

Today, we are making history. This is the beginning of the first-ever Mechanical Maintenance Committee (MMC) Conference. The MMC seed was planted about three years ago at the AMC open forum with an Air France question: Why can't we have a Mechanical Maintenance Conference? Avionics engineers save time and effort by cooperation; why can't the mechanical engineers have a similar conference?

Here we are. The MMC mission is to explore new worlds, to seek out new solutions and new designs, and to boldly go where no man has gone before.

The most charming part of the conference is that it is a proven formula. Professionals come from all over the globe, meet for two days, and solve each other's' problems. Many new ideas pop up, new friendships and great networks are created. If you do it right, you go home with a lot of business cards and many great ideas to save money for your company.

The formula has been working for the 68 years of the AMC Conference. Now and then, we improve and renew it.

The name of today's game is intellectual property. But the conference formula for you, my fellow mechanical engineers, is for free. This is, my mechanical friends, the present from avionics engineers to mechanical engineers. We all are working hard to keep the work fleet flying.

In the first place, I would like to tell you that I am privileged to be here with you sharing a passion for aviation. Honestly, when it comes to technique, aircraft performance, maintenance, and new developments, we share the same passion. There are no borders to keep us apart. We are cooperating and keeping the work fleet flying and under control. We are the engineers.

Avionics engineers are working with components that perform similar functions in mechanical systems. What we call diode, you call check valve. What we call transistor, you call valve. What we call high voltage, you call high pressure. What we call current, you call flow. We all are engineers, and we are a special breed. I am proud of it.

A few words about engineers. The definition of engineer varies in different countries. In the US and Canada, engineering is defined as "a regulated profession whose practice and practitioners are licensed and governed by law." In some English-speaking countries, engineering is seen as a dry, uninteresting field in popular culture, and the domain of nerds. For example, the cartoon character Dilbert is an engineer. Several Star Trek characters are engineers. Q in James Bond movies is an engineer.

The difficulty in increasing public awareness of the profession is that average people do not have personal dealings with engineers, even though they benefit from their work every day. By contrast, it is common to visit a doctor at least once a year, the accountant at tax times, the pharmacist for drugs, and occasionally even a lawyer. It is not common to visit an engineer. In companies and other organizations, there is a tendency to undervalue people with advanced technological and scientific skills compared to celebrities, fashion designers, sports men, entertainers, and managers.

Engineers develop new technological solutions. They engineer the design process. The responsibility of the engineer may include: defining problems, conducting and narrowing research, analyzing criteria, finding and analyzing solutions, and making decisions. Much of engineers' time is spent on research, locating, applying, and transferring information.

Aircraft systems are complex. LRUs are complex. Software is complex. Hardware is complex. The whole aircraft is extremely complex, having managed to overcome all issues to make it fly.

Thank you, mechanical engineers. Thank you, suppliers and vendors in the room, for delivering us a complex machine fit to fly. Thank you, all other engineers in the middle section (in this room, front section), for keeping them flying. Remember: the majority of useful aircraft life is in the hands of the operator. Together, mechanical engineers and electronic engineers are keeping our fleets in good shape. They, airframers, and OEMs spend a few years designing and building the aircraft. We on the operator side spend 20 or more years flying and maintaining them. Of course, there is interaction between all of us, but it is the operators' responsibility to do extremely good maintenance.

But there are challenges. Let's mention a few of them.

Integration of aircraft systems is the biggest step in a good direction, but also the biggest puzzle, if you are not used to working with super-integrated systems. Troubleshooting is impossible without at least a laptop. Modifications are impossible without aircraft manufacturers because they control the interfaces and software configurations. The task of the system integrator is extremely difficult.

Another challenge is that repair shops are in a big dilemma. The equipment installed in the aircraft is extremely reliable and is flying endlessly. Each LRU, even mechanical, has software inside. To test LRUs in the shop and do Level 3 repairs as we desire, the shop must invest in a lot of cash for testing components and dedicated software. The dilemma is high test equipment cost, low failure rate, and no flow of LRUs. Sometimes you invest a lot of cash for test setup that is used a few times a year. You need more than 100 aircraft in your fleet to be cost-effective. It is obvious that we have done an excellent job designing such superb LRUs that are flying forever.

Another challenge is that we have mentioned our friends from the legal department. It looks like they are trying to take control over our activities. Contracts are extremely important. Lawyers have managed to change our industry. OEMs are contracting the sub-suppliers and airlines. Airframers are contracting their vendors and sub-suppliers. Flight simulator manufacturers are contracting airframers and OEMs. Engine manufacturers are contracting OEMs, sub-suppliers, and airframers, and they are contracting airlines and MROs. It is important to have a contract, but it is also very complex, especially if you are not a lawyer.

Besides contracts, there are also export licenses, nondisclosure agreements, end user agreements, designee letters, delegation letters, authorization letters, and ITAR licenses. It looks like lawyers have started to control our lives.

But there is a catch: they need engineers to establish the scope of every contract. The lawyer knows perfectly how to define clauses about warranties, late deliveries, payments, delivery times, exchange points, and penalties. But they know very little about the scope of the contract. Here is the opening for us engineers.

Do you remember the story about going to the doctor, the pharmacist, and the lawyer? Well, here is the new standard expression: going to the engineer. Engineers more than ever provide data for each scope of contract. You as an engineer must be consulted; otherwise, the contract is not worth signing and the lifecycle costs will explode.

Another challenge is for repair shops: independent MROs or airline-owned MROs. There is deadly competition in the aftermarket. That is the reason airlines that are traditionally doing repairs in the aftermarket are losing big pieces of the pie each year. Can you imagine how difficult it is to step into component maintenance for an extremely reliable and expensive component, which can be tested only on an extremely expensive test set?

Challenge number five is safety. Safety is the biggest challenge in aviation. We all strive to have the same amount of takeoffs and landings. Each year breaks a record of safety. We are all here contributing to that figure. We can be proud of making the numbers and managing and improving the target level of safety. That is not easy. Our fleets have flown more miles per year than ever before. Aluminum and plastic birds (or should I say composite birds) have extremely high utilization and are hanging in the air for many hours. Due to our extremely high engineering skills, they managed to lower the number of fatalities. This is probably the biggest challenge of all, and we are successful in that.

Last but not least, the challenge of lifting up the image of the engineer. It is good to know who we are. A Google hit says, "an engineer is someone who solves a problem you did not know you had, in a way you do not understand."

Some also say you are an engineer if:

• You take a cruise so you can go on personal tour to engine room of the cruiser
• You see a good design and still have to change it.
• Your spouse hasn't the foggiest idea of what you do at work
• You try to repair a $5 bicycle pump

But who are we really?
No engineer can walk away from an unsolved problem until it is solved. No illness or destruction is sufficient to get the engineer off the case. These types of challenges quickly become a personal battle between the engineer and the laws of nature. Engineers will go without food and hygiene for days to solve the problem, and when they succeed in solving the problem, they will experience an ego rush that is better than sex.

I see that nobody is snoring, and that is always a good sign. Let us hammer our first MMC Conference!

MMC Closing Remarks by Marijan Jozic,

Just a few words about statistics. I am sure you are interested to know about the statistics of this conference before we slam the hammer for the last time.

We had 185 people registered and present at the conference, 11 suppliers had a table in the coffee area, 152 questions, and 25 open items, which is 15%. We had 16 success stories, which is about 10%. The first day, we handled 66 questions, and today, 86 questions (good job, Anand Moorthy [American Airlines]). We had one symposium in those two days.

It went very smooth. We will not need a steering committee meeting afterwards. Steering committee members, you can also relax and have a cup of coffee after your hard work.

And now, one more slam to close the conference, and applaud for yourselves.

This Happened on the Way to Cleveland

(and illustrates equivalency, acceptable operation and accessibility)

Last month, the very first Mechanical Maintenance Committee (MMC) held the first MMC Conference in Cleveland, Ohio. The MMC concept seed was planted 3 years ago at the 2014 AMC with an Air France question:

"Why can't we have a Mechanical Maintenance Committee?"

Avionics engineers save money by collaboration and cooperation at the AMC Conference; Mechanical engineers need a similar conference! Many people responded quite positively. The AMC Steering Group spent an entire day planning and executing a tabletop exercise for the MMC Conference. Could we duplicate the success of the annual AMC Conference? In Dublin, Ireland we had a lot of discussions about the need, risk management, and setup for the MMC Conference. The only certainty was the name: Mechanical Maintenance Committee (MMC). We decided to organize an exploratory meeting and invite airframers and OEMs to the next AMC Steering Group meeting, which was held in Long Beach, California. Again, there were many positive voices at the meeting. Some OEMs and airframers could not attend the meeting, but they sent strong letters of support. The MMC seeks to explore new opportunities, uncover new solutions and designs, and boldly go where no man has gone before. After two years of thinking and rethinking, we decided to launch the MMC Conference. At that time, I offhandedly suggested us to go to Cleveland, Ohio because it was a central location in North America, and there are a lot of OEMs with facilities near Cleveland. It was just a remark, but just a few weeks later I got a note from the MMC Executive Secretary, Sam Buckwalter, that he has actually found a favorable venue and hotel. After he showed me a picture of the meeting room I had that "AHA!" moment. All the bits and pieces of the MMC puzzle were becoming a sharp picture. Once I saw a picture of the meeting room I got the "all will be OK" feeling and started to prepare an opening speech.

The week before the conference, Smitty and Sam were packing their equipment to execute the conference. This time was different because they did not have to ship amplifiers, cables, microphones, or video devices. The long check list was not required because in the meeting room in Cleveland we had it all. But did we? That was a Human Factor "Dirty Dozen" moment: Workplace practices develop over time, through experience, and often under the influence of a specific workplace culture. It is important to understand that most Norms have not been designed to meet all circumstances, and therefore are not adequately tested against potential threats.

In our case the situation this is as follows: For any conference there is a packing list. Smitty and Sam would normally follow the list and bring all stuff to the conference. This time we had a deviation from the norm, paired with Sam's and Smitty's excitement and the pressure of the first MMC Conference. When they arrived in Cleveland, we found out that there was no cowbell or gavel shipped from ARINC IA. These two items are absolutely and totally essential for any conference. Without these, we cannot call people to the meeting room and we cannot open the conference without a gavel.

Luckily, we had the MMC Conference Reception the evening before and we met Mr. and Mrs. Arnold, who live in Cleveland. I happened to mention the problem to Mrs. Arnold and she immediately texted her husband to come over to see us. Mr. John Arnold is a manager at Testek, and he offered to help. He promised to come next day with a hammer and a bell.

The meeting room was full of engineers. Just 5 minutes before the opening, Mr. Arnold came in and proudly handed me a hammer (not a wooden gavel). It is mechanical conference; therefore, the hammer was more suitable. It was a fantastic hammer from Mr. Arnold's garage. The hammer can be considered "an alternative means of compliance" for the gavel. He also handed us a socalled "wedding bell," which is also considered equivalent tooling and is compliant with our prescribed bell per ARINC 668 standard. Mr. and Mrs. Arnold saved the conference The second day of the conference, there was a small modification in our equipment. Mr. Ron Parpart from Rockwell Collins came with a PMA part which could be considered as drop-in replacement for Mr. and Mrs.

166

Arnold's wedding bell. The drop-in replacement was compliant in form, fit, and function.It was an improved model of the bell, which easily emitted the 85 dB of ringing sound. Compared to the wedding bell, it emitted 20 dB more. The modification was accepted by MMC authorities because it fit better in the MMC environment. The average age of mechanical engineers is slightly higher than that of avionics engineers. Their ears are less sensitive to sound waves; therefore, those 20 dB were more suitable for mechanical engineers' ears.

To remind you about compliance with USA Federal Regulations, here is the FAR check list:

25.1309 Equipment, systems, and installations. (a) The equipment, systems, and installations whose functioning is required by this subchapter, must be designed to ensure that they perform their intended functions under any foreseeable operating condition.

In our case, the hammer and the bell were fully compliant with 25.1309.

25.1322 Crew alerting. (c) Warning and caution alerts must: (2) Provide timely attention-getting cues through at least two different senses by a combination of aural, visual indications.

In our case, the hammer and the bell were fully compliant with 25.1322. Bell and hammer alerts provided aural and visual indication.

25.1411 General. (a) Accessibility. Required equipment to be used by the crew must be readily accessible.

In our case, the hammer and the bell were accessible all the time at the conference. It is evident that engineers can fix problems. The hammer and bell are a simple example of collaboration and fixing the unexpected problem.

At the inaugural MMC Conference, we worked together to solve 151 technical items. During the two days of hard work, we had an 82% success rate and received commitments from airframers and OEMs to solve the other items over the next 12 months.

The story of Three Knows

AMC 2017 Opening Remarks in Milwaukee

My fellow engineers!

On behalf of the AMC Steering Group:
Welcome to the AMC/AEEC meeting!
Welcome to Milwaukee! My name is Marijan Jozic,
And I am the Chairman of the AMC conference.

First of all, I would like to remind you that all that we do — started about 115 years ago, when two bicycle makers known as the Wright Brothers built the Wright Flyer and did the first flight at Kitty Hawk, North Carolina.
Not long after that, approximately 25 years later, General Jimmy Doolittle became the first pilot to take off, fly and land an airplane using instruments alone, without a view outside the cockpit. That was in 1929.
Now, approximately 85 years later, we are in the A350/787 era. In all those years, somebody had to design and maintain those aircraft. And we know who somebody is: The engineers. Every generation of engineers had to design or learn new things and process a gigantic quantity of data.
But that could only be possible because of **Three Knows**.

Let me introduce you to Three Knows: The first Know is **Know-How**. Know-How can be learned to a certain extent. Most of the knowledge can be learned from courses: self-study of CMM, AMM, SRM and many other books and drawings. Normally speaking, the engineer will be able to maintain and even modify aircraft and keep it flying for 40 years. Know-How is transferred from engineer to engineer and all the details are somewhere in the books. But more important, is the second Know.

It is **Know-Why!**

Know-Why is usually in the heads of people. Only those who Know-Why also know where it is noted or recorded. The ability to find it, is crucial. Let me give you two trivial examples.

The first example is the time needed to evacuate the aircraft. Every one of us knows, that in case of an emergency, the aircraft must be evacuated within 90 seconds. The valid question is:

Why 90 seconds? Why not 60, or maybe 120?

Based on the rule of 90 seconds, it is calculated how many doors you need in the fuselage. Should it be 10 or 8 or maybe 6?

Try to find out why it was decided that 90 seconds should be the regulation! It is not that easy.

The second example is the duration of power interrupt of 200ms. In many ARINC documents, is mentioned the 200ms rule. Why 200ms? Why not 300ms or 100ms?

A power supply which can maintain voltage for 300ms after power interrupt can be designed - as well as a power supply which can, during power interrupt, maintain voltage for 100ms. But **Why** was chosen for 200ms? It will be interesting to **Know-Why!**

Check for yourself if you know the answer to the second example.

If you are designing a new aircraft, system or LRU, it is crucial to know why regulations and requirements are such as they are. Even if you are re-designing the aircraft (737MAX or A320NEOS), it is crucial to **Know Why** some things are designed the way they are designed.

In the lifetime of 737 and A320, there will be upgrades where design engineers will scratch their heads and wonder WHY?

So, now it is the right moment to meet the Know number three.
It is **Know-Where**.

Know-Where means: Where can it be found?
Know-Where is of paramount importance. When you forget the past, you are doomed to do it again. Here is also the moment that we have to discuss the way in which we have stored data. If you can't find **Why,** you are doomed to re-design it again, to re-invent something which was invented many years ago. The way we have stored it now, enables us to search using keywords and computer - which is a huge improvement of today's filing.

But we still have a limited amount of people with crucial knowledge who possess the information. But if they are gone (due to retirement, winning the lottery etc) nobody can find their documents to **Know-Why** something is done the way it is done.

Fortunately, thanks to computers access to information is much easier nowadays. It took some time to get so far. This is however an essential point of our discussion. The flux of people going through airline engineering departments is so large, that over a period of time, certain essential safety, knowledge & experience will slowly be lost. We know that in the good-old-times engineers stayed in a department for 4 years and then they said, "Well, now I know the business. Let's have some engineering fun!" And they stay for another 5 to 6 years at the same job. I can see that young engineers now days move from position to position every 2 to 3 years.
Therefore, it is important to **Know-Where** to find the information.
I am certain that **Know-How, Know-Why** and **Know-Where** can be preserved using the internet and intranet. A healthy engineering environment is a good team of seniors & juniors. A blend of knowledge from the past, honored skills over 10-20 or more years, combined with fast-learning juniors.

The culture of transferring **Know-How, Know-Why** and **Know-Where** should be established everywhere.
I see that you are still with me! And this is always a good sign.
So let's fire up our great conference. Bang!

Volare Award winners (old and new)

AMC/MMC Steering Commitee

Roger S. Goldberg Award

Presented at the AMC Opening Session by Ted McFann AMC Vice Chairman and Dean Conner, United Airlines

Marijan Jozic, KLM Royal Dutch Airlines; 2017 Roger S. Goldberg Award Recipient

Hello. I am Ted McFann, Vice Chairperson of the AMC Steering Committee. As the Vice Chairperson, I have the honor and privilege of annually presenting the Roger S. Goldberg Award to a deserving member of the avionics maintenance community. Unfortunately, this year, the FedEx team and I are not able to be present for the AMC. Regardless, I did not want to pass up my responsibilities, and thanks to the FedEx communications team and fellow AMC Steering Committee member, Dean Conner, I am with you in mind and spirit.

The Roger S. Goldberg Award was created to recognize individuals who, like Roger, exhibit the characteristics of having extraordinary ideas, outstanding service to the avionics maintenance community, endless passion for making avionics better, and are a representative of the true heart and soul of the AMC.

This year's winner earned a Bachelor's of Science degree in Aerospace Technique in 1978. In 1986, he earned a second Bachelor's of Science degree in Electronics and Telecommunications. He has shared his avionics knowledge with the industry as an author of two books and numerous articles in Plane Talk magazine. His writing stimulates us to ask: are we doing the right thing, or is there another way? In one Plane Talk article, Avionics Master Plan, he struck a significant chord in all of us to keep every airplane flying from the past to the present, and from the present into the future. Last year, he campaigned to make avionics great again. He has chaired three AMC Working Groups: the SCEA, MSI, and OMG. He has been the AMC Chairperson for the past five years. Until this moment, he was not advised he was receiving the award this year.

It is my great honor to present the Roger S. Goldberg Award to Mr. Marijan Jozic of KLM Royal Dutch Airlines.

Mr. Jozic's remarks:
Thank you very much. I am speechless, because we had several AMC Steering Committee meetings and they were keeping that a secret from me. At the last meeting we even said, "Well, there will be no award this time, so Dean will step up and say, 'This year we do not have any award.'"

I am speechless. I do not know what to say. That has not happened many times for me.

Thank you very much. Instead of hugging you all, I will hug Dean.

Tribute to John Barker

The Canada Music Company

The AMC is obviously the mother of all conferences. Many times in the past I have touched on that theme and tried to explain the importance of the AMC conference. Even more, I have tried to explain that this conference is not just a conference. It is the people, the attendees, that are making it the greatest conference of all conferences. We are very fortunate that the newest engineers attend and that it is a very special mix of engineers from OEMs, airframers, and operators. They attend because they care about aviation and safety, as well as lowering costs.

That engineering mix at the AMC are collectively making important decisions and creating standards which are used by others. These others who don't even realize that we, the AMC people, are making brilliant and important decisions for them. Whether they like it or not, they are following us. Those of us that participate in the AMC are the industry leaders, and I am happy to belong to this group of leaders and not remain as an industry follower. If you are a follower, you are, per definition, always one step behind. The people of AMC are a special breed, the breed which feels comfortable to lead. Because we all are very involved and have known each other for many years, a friendship is automatically created. We see each other maybe once a year at the conference, but we know that our buddies are always available if we need some help. At any moment of the day, and on any day of the year, we can call our friends from another continent and they will be there for us.

John Barker retired from CMC in 2017. Tribute to John Barker—The Canada Music Company Alas, each year some of these good friends are retiring. After years of being leaders and great participants at the AMC conferences, each proud engineer is able to savor the sweet success of retirement. Sometimes, we may not know that someone has retired. Some of them just disappear and start to enjoy the retirement.

Others have great friends who celebrate our years of working together and boldly recognize our upcoming retirement. And in 2017, one of those AMC and industry buddies is John Barker. My next story is about him. He stands alone as an aviation professional, and just happens to be a great guitar player. Or, it may be argued, he is a great guitar player who has a day job as a brilliant engineer. For many of the past AMC and AEEC conferences, John and Phil Moylan have entertained us at the CMC Electronics suite and many of us have had the opportunity to sing with them. It was a time-honored tradition. I fondly recall the moments at a few conferences when I first entered the CMC suite. John and Phil immediately shouted to me, "Hi Marian! Let's do Proud Mary." And we sang Proud Mary and many of great songs together! John Barker, It was my privilege to know you and to sing with you.

From Small Beginnings Come Great Things

This is a story about the beginning of a new ARINC Industry Activities conference, which is called MMC – the Mechanical Maintenance Conference. Let me tell you how this conference started. In the last couple of years, avionics engineers were reporting that their buddies working on mechanical systems were jealous because they didn't have a conference like the AMC. Then, Air France asked two times in two different AMC conferences about the possibility of a conference dedicated to mechanical systems. We discussed it in the AMC open forum and then later in the AMC Steering Group meeting. We were scratching our heads and asking ourselves if and why a mechanical conference was a good idea. Then we decided to add one more day to our AMC Steering Group meeting in October 2015 to thoroughly discuss what we named the MMC. We figured out that the beginning is always the hardest. Nevertheless, we took up the challenge. The first step is to organize an exploratory meeting and invite professionals in engineering and maintenance on aircraft mechanical systems. This was no small undertaking as we knew the potential impact to the industry.

The biggest challenge will be to attract a sufficient number of airlines and their respective engineering representatives to participate in the conference. We all know from our experiences at the AMC Conferences: if airlines attend, the suppliers will attend too. So the MMC exploratory meeting took place in Seal Beach, Florida, and we invited a few airlines, airframers, and equipment suppliers to attend. After the MMC Steering Group meeting, we met with suppliers and airframers. Airbus announced that they would support such a conference and after just a few hours of meeting, we all felt that we were on the good path. The decision was made – a new conference would be organized by ARINC Industry Activities. We are aware that the beginning is the most important part of the work. Therefore, we decided to systematically map the challenges and work through them, just as the AMC planning committee does every year. At this moment, we have passed the point of no return. The conference is on the calendar and there is no looking back. .

Let me reveal the details. The hotel contract is set. It is at the InterContinental Hotel Cleveland in Cleveland, Ohio. The dates are

November 7-9, 2017. On ARINC's website, there is information about the MMC and you can find it under the AMC Tab. Tuesday, November 7th will be the exhibit reception Expo and the technical conference will be Wednesday and Thursday.

1. Inform engineers from mechanical engineering groups about the MMC. Tell them about your AMC success stories and how avionics engineers solve problems in the open forum. In Milwaukee we had quite a few success stories that went above and beyond simply answering the airline's technical questions. We solved many complex problems, but approximately 20% of the problems are solved with extraordinary cooperation between suppliers, operators, and airframers. Those success stories are extra important because they would not be solved without the AMC. That is exactly the spirit of the AMC and we would like to pass this spirit on to the MMC.

2. Let your mechanical engineering friends read the article in this Plane Talk from my buddy Ted McFann from FedEx. The article is about the justification of attending MMC and about the questions and answers you can expect when you go to your boss and ask for approval for the conference. Rest assured that we all go through that step. Here it is important that you not lose faith. If your boss says "No!" then that only means that your story is not good enough. Tell him that you will come back with a more thorough justification. If you need the invitation letter from ARINC Industry Activities, please contact Sam Buckwalter at sam.buckwalter@sae-itc.org.

3. On the ARINC Industry Activities website, there is information on the upcoming MMC. Possibly the most important link is how you can submit technical questions about a chronic or difficult mechanical issue you might have. Also, please do not make it a surprise for the airframers or equipment suppliers. Before you submit a question to the MMC program, you should let all concerned parties know that you are submitting the question to MMC. If supplier does not respond, go ahead and submit the question. The point is to be fair, because in the spirit of the MMC, you don't want the supplier to say "I did not hear about that before the conference." Give them a fair chance to show cooperation. The point is not to shame anyone, but to have a collaborative discussion about solving our industry's challenges..

4. The open forum discussions will be logically categorized by ATA Chapters as well as three additional sections: Maintenance philosophy Testing systems Ground Servicing Equipment and Tooling

5. The conference is scheduled for two days and a big portion of time we will spend in discussing your questions in the open forum. The sessions are led by a moderator. Every question will be presented on large projection screens. The long form of the question can be read from the MMC Program. Meanwhile, the moderator will summarize and invite comments from airlines. Any airline or operator can add important information, or simply speak up to show their support for their fellow airlines. You can expect comments such as: "We have experienced the same problem and we support ACME airlines with this issue". The moderator will then ask for supplier's comments. The supplier can then explain their plan for remedying the problem, or summarize how they assisted the airline in the problem's resolution. The last one to speak is the airframers (Boeing, Airbus, Bombardier, Embraer, etc.). Most questions have positive outcomes and the answer is shared across all airlines. This is the biggest value – sharing engineering and maintenance experiences that make the industry safer and more efficient. These are truly "success stories".

6. Besides the open forum, we will have a symposium session dedicated to an important industry topic or challenge. This year we have chosen Parts Manufacturing Approval (PMA). We will strive to have presentations from different perspectives: airlines, suppliers, and regulatory authorities. Following the symposium presentations will be a question and answer period from the conference floor. This is exciting to watch – knowledge and experiences of industry experts is transferred to the entire audience of engineering and maintenance professionals.

7. The MMC Program will be published approximately 6 weeks before the conference. The schedule of events is roughly the same as AMC: Opening Session – MMC Chairman, Keynote Speech, ARINC IA briefing, etc. Open forum discussion separated by a lunch where attendees can visit expo booths Symposiums are after the afternoon coffee break.

8. The MMC will welcome airlines and ARINC IA sponsors to the Tuesday evening reception. The MMC attendees will have an opportunity to

network and peruse equipment suppliers' booths during the 2 hour event. Displays are sure to include PMA parts, test equipment, GSE, etc.

9. The second day the MMC will continue with the open forum discussion items, and will conclude with a discussion and survey about the MMC. This is an important moment – we will need feedback to steer the MMC conference's future towards success.

10. Behind the scenes at the conference, the MMC Steering Group will meet each day. This first conference has a steering group comprised of airline avionics professionals. But this is only due to their experience with the conference format and execution. An election will be held in the open forum to elect airlines to the MMC Steering Group. The group will be comprised of 11 representatives, preferably representatives of several regions. So think about it right now and announce in timely manner if you want to be a part of the steering committee as a member. If you feel that you want to do more, step up and say that you want to be a candidate for Chairman or Vice Chairman. I cannot stress enough how important a conference steering group can be. There is one important statement to say at this point: "The more you bring into the MMC conference, the more you gain."

Every year AMC saves time and money for the airlines by finding solutions to avionics issues. We know that the MMC conference can do the same. The key is cooperation between airlines, airframers, and equipment suppliers. That is how it works.

This was MMC in a nutshell, from the perspective of AMC chairman. There are some minor details. It is also important to note who, how, and what supports the AMC and MMC in their mission.

ARINC Industry Activities organizes aviation committees including the:

- AEEC Airlines Electronic Engineering Committee
- AMC Avionics Maintenance Conference
- FSEMC Flight Simulator Engineering and Maintenance Conference

The ARINC IA staff provides engineering support subcommittees and working groups in the preparation of ARINC Standards for their respective committee (AEEC, AMC, FSEMC). This team of project engineers and editorial assistants also prepare, execute, and report on the large conferences each year.

ARINC IA (ARINC Industry Activities) is doing three important things for Aviation:

Conferences: AMC | AEEC, FSEMC, EFB Users Forum, and now MMC

ARINC Standards: Maintaining over 300 aviation standards and issuing 15 or more new standards each year.

Educate: At each conference, we organize numerous seminars to spread knowledge and experience to help our members understand the newest development in aviation maintenance

If you are reading this, please forward it to your mechanical engineering and maintenance colleagues. Ask them to subscribe to the PlaneTalk mailing list. PlaneTalk will continue to have information specific to the MMC, so pass along the information with a wide distribution. Plane Talk is the AMC's magazine with a very long tradition. It is also the AMC's platform to communicate with our members between conferences and meetings. After 68 years of providing value through the AMC Conferences, we know how to engage and hold people's attention. We want to share this wisdom with the mechanical side of aviation. For 3 days in November, the MMC will duplicate the success story of the AMC. As you know, communication is key to the success of any organization, and the AMC is living proof of this. By November 10th, the MMC will prove it as well. If you dream it, and if you build it – success will follow!

Everything you Wanted to Know about Repairs

(but were afraid to ask)

Aviation has always been a special business. Modern aviation is even more special. As we all know, aviation is very well organized and regulated. Especially the regulated part. When we (or a pilot) discovers that an LRU is not performing, the technician removes the LRU from the aircraft and sends it to the shop. The shop in an aviation environment is known as a Part 145 organization. That means that the LRU must be repaired according to Component Maintenance Manual (CMM). Here is the first dilemma. We are talking about "repairs" and then using a Component Maintenance Manual. So, is it a repair or simply maintenance? Well, what we call repair is actually maintenance. By means of the CMM, we restore the LRU to its original approved and working condition. If you send the LRU to the shop and you receive it with a regulatory certificate (FAA 8130 or EASA F1) you would normally say it is repaired in the shop. And that makes it a bit confusing because the shop actually restored it. Even if a piece part is replaced by a new working part, it is still restoration.

When we talk about real repairs, it is something different than restoration or changing parts. It is an action what we call: "fill and drill." For example! Part of a printed circuit board is burned due to high current through a broken resistor. The technician cuts the burned Printed Circuit Board (PCB) area, fills and glues it with PCB material, drills holes, installs the new resistor, applies conformal coating, and then tests the unit. That is the repair.

The repair is sometimes not described in the CMM but in the Standard Practices Manual. Every technician must be aware of those standard practices and the Standard Practices manual must be available in the shop and revision-controlled. Every component supplier has it and they use it every day. You as a Maintenance Repair Organization (MRO) can decide to issue your own Standard Practices Manual and make it part of your quality system. In that case, you will be responsible for its content and revision service.

There is, of course, more than one way to skin a cat. Some of us decided to issue additional pages and add them to CMM. Some call these pink pages, some blue, and some green. Color does not matter as long as you can identify them and see that they are different than CMM pages. By means of those additional pages, you can add information to your CMM. It can be just something trivial and easy. For example:

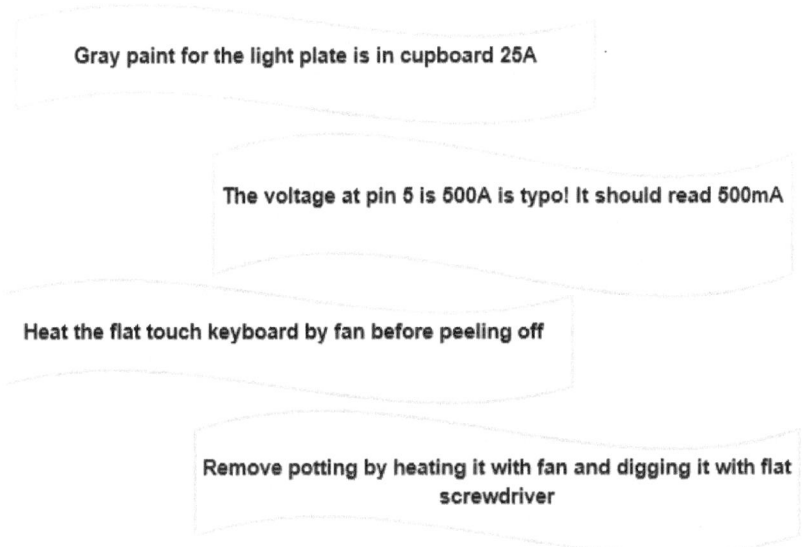

Gray paint for the light plate is in cupboard 25A

The voltage at pin 5 is 500A is typo! It should read 500mA

Heat the flat touch keyboard by fan before peeling off

Remove potting by heating it with fan and digging it with flat screwdriver

As you can see such a system opens a whole new area of repairs. Suddenly, you can do much more as long as: The repair is part of your quality system. Your repair is within the specification of the LRU. Your actions are prescribed and part of your quality system

The easiest way of performing maintenance is just to replace PCBs or the whole assembly or module. You will immediately find out that you are throwing money down the sewer. This is the absolute most expensive method of maintenance. Therefore, to work effective you should accomplish maintenance to the smallest degree possible. Call it repair if you want. Level 3 of maintenance is described in several ARINC standards. ARINC Report 663: Industry Guide for Component Test Development and Management defines Component Maintenance Level (CML) 3.

> *Maintenance operations with an end result of restoring a LRU or a subassembly to serviceability. Maintenance includes repair of the LRU and or its subassemblies by any and all repair processes, including but not limited to, replacement of defective Basic Parts such as processor chips, transistors, or chassis mounted parts. CML-3 embodies those activities necessary to: Fault isolate to the Basic Part level Replace/repair such parts Return the LRU or subassembly to service This level (3) includes all programming, calibrations, alignments, tuning, etc., necessary to return the LRU/SRU to service.*

This is good stuff to know. Actually, technicians of every Part 145 organization should dream about it every night and use it in everyday business.

One step further is a repair which is not classified as standard practice nor described in the CMM. You might say it is not repairable! But! Here is one important BUT! This "but" is about money, honey! It can happen that the part is very expensive or it can be a case where that part is relatively inexpensive but fails a lot.

At a recent AMC, I bumped into an engineer who told me: "I love parts that are very expensive or are cheap but failing very often." Those parts are the target for investigation. Here are the questions to ask in an investigation:

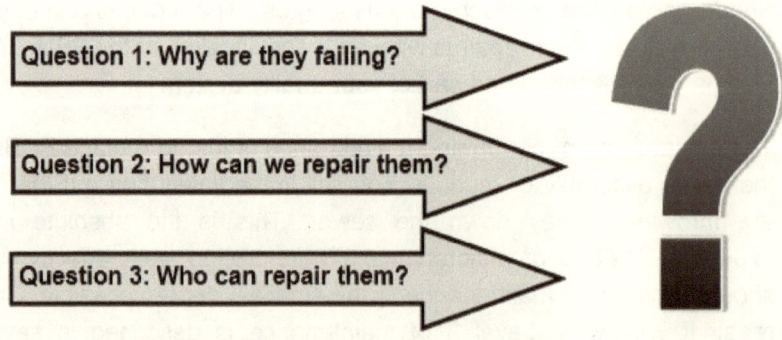

Question 1: Why are they failing?

Question 2: How can we repair them?

Question 3: Who can repair them?

You might have noticed that I purposely did not ask: Can it be repaired? That is because I am convinced that everything can be repaired. It all depends on how much are you planning to spend for repair and how expensive the new part is.

Here's a nice story: At home, I have a different problem. I am convinced that many engineers have the same problem: I can fix everything. Well, in last 35 years, I have fixed everything. The other side of the story is that my wife blames me that we only have old stuff. Even if the parts are obsolete, I simply redesign it and it can be used again for many years. She hates our 35-yearold washing machine (and dishwasher, and vacuum cleaner, and mixer, and hairdryer, and radio, and...) . Therefore, I am convinced that everything can be repaired. But there is a limit. If the repair is more expensive than a new LRU, scrap it! But make sure that you are certain that repair cannot be done cheaper. At home is different story. I can spend 2 days repairing our dish washer simply because I enjoy it. But if you add the engineer's hours of labor into equation, you can easily buy a new dishwasher. That example does not make sense.

Back to those three questions. If you cannot repair the part and you still think that it can be fixed, submit the question to the AMC or MMC and simply ask the other 400 engineers in the room. You can also come to AMC or MMC and check very thoroughly if somebody can either repair or design a repair for you. The best place to talk about that is during the AAI reception. Everybody is there and there is an immense amount of brain power in the area. The chances are good that one of those few hundred engineers can give you the golden secret. And that will save you a lot of money.

My golden secret is actually very similar to when you go shopping in Amsterdam: You can buy the best stuff in the smallest shops. One of my bean counters used to say:

 "The money lays on the ground! You just have to bend over and take it." That works also with repairs in aviation. If you are too lazy to bend, it is too bad for you!

Pitot Probes — A Simple Fix

Picture this: June 17, 1994, the pitot probe users group meeting just finished, and we were enjoying the evening. On the TV was the O.J. Simpson chase, you know - the white Ford Bronco, Los Angeles, police chase, etc. The rest became history. I tell this story because it positively marks the date of the pitot probe users group meeting.

Te meeting was the result of numerous complaints about the quality of pitot probes and pitot/static probes. Airlines were explaining that probes were failing at the most unexpected moments and there was no indication that a probe was degrading. Probes were failing with low amount of hours in use, so the manufacturer decided to organize a meeting to discuss the problems. This was one of the most constructive meetings at a manufacturer that I have ever had. We discussed the probe materials, the way they built it, and the way they tested it. There was a lot of positive energy. We had excellent deliberations and by day two, it was time to reveal some data. First, it was announced that probes are made of beryllium-copper and that they suffer from corrosion.

If an airplane flies above industrial areas, acidic rain penetrates the probe and corrodes the material. At a certain moment, a small crack occurs in the shield of the heater. If the airplane is flying and the heater is on, nothing happens. However, when you park the aircraft at the gate in the rain, the water will penetrate through the crack due to capillary action. Capillary action (sometimes capillarity, capillary motion, or

186

wicking) is the ability of a liquid to flow in narrow spaces without the assistance of, or even in opposition to, external forces like gravity. It occurs because of inter-molecular forces between the liquid and surrounding solid surfaces. If the diameter of the tube is sufficiently small, then the combination of surface tension (which is caused by cohesion within the liquid) and adhesive forces between the liquid and container wall act to propel the liquid. When the aircraft is parked, the liquid will accumulate in the small tube which holds the insulation and heater element. Over time, the insulation material becomes soaked.

Then the air crew arrives. They start up the airplane, followed by push back and starting the engines. When they get clearance to taxi, they switch on the pitot probe heater. And a minute later, they see the pitot probe fail indication. What in the world happened? The water surrounds the heater element and when the heater is activated, the water starts to evaporate but it can't easily escape through the little crack. The small explosion makes the hole bigger and then electrically shorts the heater. The air crew in the cockpit sees the resultant red light.

That explains the most common case of a pitot heat fail. It is a repetitive failure. The problem is that it occurs at the worst moment, directly after push back. It causes an abundance of grief. The resulting fix takes a lot of resources, including:

A. Pilot returning to the gate
B. Ground crew fetching the probe from the store
C. Replacing the probe
D. Performing the leak check of the system (if no quick disconnects)
E. Aircraft departing again

The delay is typically 90-180 minutes, depending if a leak check is required. Nobody has time for that. So, in 1994, 23 years ago, we knew the cause of the problem. At that meeting, we defined possible preventative actions:

a) Nickel plated probes
b) Integral heater
c) Non-metallic probe
d) Dual heater

Commentary: These days, we would add a 3D printed probe, but we didn't even dream about 3D printing in 1994.

The end of the story is that the probe manufacturer decided to roll out the new nickel-plated probe. That was quick and easy. The probes were then making up to 4000 more hours on-wing. Of course, the amount of complaints also diminished. The manufacturer never designed and built a non-metallic probe because the probe design was supposed to be "fit and forget". Conversely, the ceramic material was resistant against corrosion, the probe could be heated through nonmetallic material and without using the wire heater. It was a promising design, but an absolute killer of aftermarket revenue. The manufacturer also never launched a design with an integral heater. This design was found to be difficult to accomplish. The strut could be heated but not the tube. This was, ultimately, the problem that made the design unacceptable. In theory and expected practical use, a dual heater design would be an ideal solution. When one heater failed, we would know that we should replace the probe after the flight and there would be no failure at taxi way. For this solution, we would need the airframer to install some additional wiring, a change in the indication lights, and to certify the system. Why am I telling you this? Because I certainly became an expert in pitot tubes. I can call myself some kind of an expert because I was fighting with the probe supplier and Boeing to get the B737 pitot probe improved. The first versions had three separate copper pneumatic tubes at the bottom of the strut. During removal and installation, the probes were very often damaged by the torsion of those copper tubes. After I received a brand new probe in my office with hopeless damage of the tube by torsion, I decided to start to fight. It took me almost a year to convince the probe manufacturer and Boeing to connect those 3 tubes with a metal bar to strengthen the assembly.

I even have a letter from a Boeing Vice President announcing that they would roll out a new probe design with those 3 tubes connected. After 20 years, nobody knows why those 3 pneumatic tubes are now connected. I am the only one who knows why!

Alright, after saying all that, I can only add one thing: I don't want to be emotional and angry, but it remains painful that we still suffer from pitot probe heater failures. All other aircraft parts have improved in the last 25 years with the exception of pitot and pitot-static probes. It really is painful.

The Boeing B787 is delivering 500 gigabytes of data after every flight and we still cannot predict when a simple probe will fail. At the AMC conferences, we have asked many times: install that second simple heater. It is not a sophisticated option, but it would work. Just calculate for yourself how many resources your airline uses for new probes each year. Also, calculate for yourself the cost for an unplanned technical delay of 2-3 hours. The present design is deficient.

It is time for a change! We are able to send spaceships to the planet Pluto (is it a planet again?) and we have a couple of rovers driving around on Mars. Elon Musk designed the electric Tesla car with an autopilot. We have smartphones in our hand, which have better computing power than all computers used in the Apollo program when we put humans on the moon. Everything is smart and modern, but we still cannot obtain decent pitot probes.

It is time for change. The AMC Steering Group is going to organize an seminar and invite probe manufacturers and airframers. Would that help to get the wheels in motion? I don't know. It is time to fix the problem. Of course, there is more than one way to skin a cat. I know there is more than one OEM who is interested in changing the world. So let's cooperate and fix it once and for all. Up until now we have followed, and look what happened. It is time to take a lead.

Millennials are Smarter?
(but do not tell them that)

As time progresses, it shows that every new generation of engineers will develop and maintain a new aircraft type. We have seen it throughout the past and it will happen in the future. There is always a small overlapping period, but who cares. When I started my avionics career, the Boeing 747 classic was the newest technology. It took me some time to master it. I know that previous engineers were surprised that new guys and girls were allowed to work on it and develop modifications. But it was different time. One of the older engineers told me, "I was allowed only to look at the autopilot MCP for two years before they allowed me to push the buttons and turn the knobs."

Shortly after that conversation, I became seriously involved in bringing the B747-400 into our fleet. This airplane was state of the art at that time and now, 25 years later, I am witnessing the process of phasing out the B747 at the sunset of its life cycle. I have also seen the MD-11 type coming and going as well. To add to those huge parts of our fleet, I have seen the A310, B767, and B737-PG fleets all change and, sometimes, decommissioned. But the B787 is a different cup of tea. In the same way, the A350 is very different. At this moment, I cannot imagine what kind of aircraft the B797 or A390 will entail. I suppose the new generation of engineers will take care of that.

In the meantime, new, fresh, young engineers are entering the aviation theater. They will start at the point where generations of engineers successfully brought the aviation industry from the year 1903 to 2017. During the overlapping period, the experienced older engineers are still mingling and working with the young ones. Therefore, the young engineers can and must learn fast. Very quickly, the students will be better than their teachers.

For example, in 1970, if you were asked, "Who are the most knowledgeable people on the planet?" The answer would be: university professors, research scientists, and the occasional well-intentioned journalist. But today the most knowledgeable people on the planet have just recently learned to drive. They eat sugar cereal and do not wear a necktie. They are the millennials, those born between the late 1980s up to the year 2000. Over the next decade, 80 million retiring baby boomers (engineers and people who are now running our industry) will be replaced by 75 million millennials in the workforce. That is what almost every statistic states.

Unlike those born before humans adopted the same digital middle name (@), millennials have not suffered the hardship of having to read, remember, and recall information for exams, book reports, or wedding speeches with the same rigor and difficulty of prior generations. To millennials, the world is just a click away. They all have state-of-the-art smart phones. Their philosophy is: why memorize anything when you can search everything? What millennials lack in wisdom, they make up for in knowledge. The only difference between knowledge circa 1970 and millennial knowledge is where it is stored: rather than keep it in their heads, millennial keep it at their fingertips. Why clutter the storage capacity of the mind when you have a terabyte under your thumb? Here's why. Although our next generation leaders may be able to navigate information more readily than previous generations, this begs the question: Will they be able to create with the same capacity? I have two millennial engineers in my family, and I continue to be amazed at how they communicate. With MP3 player earbuds permanently affixed to their heads, their faces buried in the glow of their cell phones, and their thumbs pounding out text messages to friends they have never met face to face, they walk and talk. Their friends are in another country and they might never see them face to face, but they know more about them than about other millennials in the same city.

Why write things down anymore when your tablet can voice to text?

Millennials do not remember a time before computers, cell phones, or the internet. Parallel processing and multitasking are not only technology capabilities of this generation; they are a way of life. What concerns me most about this newest generation of emerging leaders is their relationship with knowledge and the pursuit of it. My concern can be summarized by the response I received from one them, in answering my question about his favorite new product, he replied, "My favorite new product is Wikipedia."

"Why?" I asked. "Because," he answered, "you don't have to think anymore."

Clearly, he is not alone. However, one of my millennials in the family once told me that he hates his ability to remember useless information.

OH, SO YOU'RE A MILLENNIAL

DO TELL ME ABOUT YOUR PLAN TO BE CEO WITHIN A YEAR

He knows for example all the dialogue from 14 episodes of The Simpsons although he saw each episode just once. The same with Star Wars and Dr. Who. He admitted that he would like much more to remember other more useful facts. Today the most wired people are the most knowledgeable people. This, however, does not make them the most creative. In fact, I would argue that they risk losing their creative capacity by not exercising their memory and attention skills. It might be a temporary effect of being wired and constantly plugged in 24/7. Speaking of aviation, those emerging engineers will be able to manage projects we started like aircraft connectivity, big data, free flight, pilotless cockpits, etc. I have no doubt that they are smart. They are a different kind of smart than baby boomers, but they are definitely smart enough to take over the controls. Creativity will come back in that environment, but possibly in a way we do not understand at this moment.

I still remember the movie Space Odyssey 2001, when the computer Hal said: "Dave, the high gain antenna will fail in 48 hours!" That is exactly what we want from our millennials. To crunch the big data and predict when our equipment will fail. I have no doubt that they will manage that, even with MP3 player earbuds permanently affixed to their heads.

Honestly, the millennials that I have met in the last couple of years are good engineers. I have no problem if they take over our jobs as we retire. The industry will be in good hands and they will be better and smarter than we are.

To Boldly Go Where
Nobody Has Gone Before

We just held the inaugural Mechanical Maintenance Committee (MMC) Conference. People sometimes say: "We are Avionics engineers! Why would we even think about mechanical maintenance?" Yes, we could say that. But saying that it does not make it exciting. We are aviation engineers and we are supposed to evaluate, collaborate, and improve. We avionics boys and girls know the power of a focused maintenance conference because we have a tradition of 68 years of the Avionics Maintenance Committee (AMC) and counting. We also know that we are not alone and that an airplane is more than just avionics systems. An airplane is just bunch of parts flying in close formation. And in today's aircraft, you will find wiring holding it all together. Almost every part of the airplane is connected somewhere by wire. Every part has sensors or microcontrollers, so hence, every system could be considered an avionics system. But avionics engineers recognize that an airplane is actually quite a bit more. There are the primary structures, empennage, engines, and mechanical systems. Engines are, of course, a different Type Certificate (TC) and a much different world. Let us put them aside, although they cannot run without electricity. Primary structures used to be made of sheet metal, which was held together by rivets.

For modern airplanes, we can put aside the composite fuselage and primary structure, which is often held together by glue. What remains as uncovered are mechanical systems. That is exactly the reason why we boldly step into this new world: the world of mechanical systems. With the initiative and supervision of the AMC, the new conference was organized. New? NEW? Well, for some of us, it is new. It is the same successful formula we are use in the AMC Conference. Airlines submit technical questions and describe their chronic problems. We publish these in a conference program and then think about it. Not only those that will attend to conference, but many engineers at the home station who work for operators, OEMs, and airframers. Eventually, at the conference we discuss each question, collaborate, and provide solutions to each issue. The issues can be something seemingly simple, or complex problems. And for every win-win successful item, we officially deem them "Success Stories!" The MMC Conference saw a lot of these in Cleveland!

The format of the AMC Conference was designed and improved by avionics engineers over the last 68 years. This makes it very stable and reliable. The airplane operators who attend the conference are lucky in several ways. In the first place, they meet fellow engineers and are able to collaborate and talk about "their" aircraft systems and their issues. Secondly, their problem formulated in the submitted question can be solved, which often brings them a great deal of savings. The next step is education. We also educate. We do fantastic seminars and this year, there was a great topic: PMA parts. Many people were telling me before the conference: "Mr. Chairman, are you sure that you want to discuss PMAs? All of the OEMs will get angry!" That question made me think. First, the answer is NO. They will not get angry, because they are aware that PMA parts are here to stay. Secondly, it does not work that way. They cannot be angry about their customers who spend millions buying their products. If a customer decides to introduce a PMA part, the OEM will not be ruined, and their customer will gain some advantage. So this is not the problem.

Now, let us go back to my thinking. Why would some individuals think that OEMs will get mad? This is very strange. It could be that there is a lot of bad publicity surrounding the PMA parts concept. Which is absolutely incorrect. From an FAA perspective, it does not matter if a PMA part is produced by an OEM (licensed PMA) or by other party (test and computation PMA). The FAA would never approve inferior parts; therefore, all doubts can fade away. All the bad publicity about Test and Computational PMA parts can be considered fake news. Finally, do not forget the legal battles about PMA parts. Many OEMs insist putting in contracts the clause: "No PMA parts are allowed on our products," meaning Test and Computational PMAs. In contract negotiations, there are always at least two parties. Therefore, it is dependent on the skill of contract negotiators if your contract will have the clause: "NO WAY PMA!"

To Boldly Go Where Nobody Has Gone Before

PMA parts are either duplicates or shown to be improvements over original parts. Crying tears of joy may well be the body's way of restoring "emotional equilibrium".

And now back to the MMC Conference, just for a moment. A very short time after the start of the conference, the engineers from OEMs, airframers, and operators all became familiar with the idea of how the conference works. The learning curve was quite short, and the operators found that open forum discussions were useful, even if we were handling questions of other airlines' colleagues. Also, the collaborations and discussions during the coffee breaks were valuable. The most exciting moment for me was that after the close of the conference, people remained in the meeting room. They formed small groups and continued with their discussions. When I saw that, I had "happy tears" in my eyes. I regret that I did not shoot a few pictures of that scene. That was a historical moment. We had at least 15 small groups of 3 to 4 people who were discussing their particular topics, exchanging business cards, and making plans for their next steps in fixing an issue. That is the real power of the MMC Conference. We were bold, and we planted the MMC seed. And that MMC seed started to grow and thrive. Hopefully, the MMC tree will be as large and healthy as the already mature AMC tree.

Welcome to the 2018 AMC in Dallas!

As the old cowboy saying goes, "Do not go where the path may lead, go instead where there is no path and leave a trail."

I would say it is also applicable to avionics engineers. Good engineers should not just follow the path. They should deviate from the path, look at the problems and challenges from a different angle, and leave a trail that should be followed.

I have said many times that we, the AMC community, are here to lead, not to follow. We are very brave and smart, and we are stipulating the way others will follow.
We, AMC engineers, are the three important pylons of Avionics Maintenance Committee. We solve problems, set the standards, and educate. Isn't that a noble duty?

Some of us are experts in certain areas. We share our knowledge in numerous seminars that we organize. That is called the AMC spirit.

Many of us have a problem and ask questions in the open forum. Some have answers and share the knowledge to fix those problems. That is called the AMC spirit.

The whole aviation industry needs guidelines and standards. We sit together, and in three meetings, set the new standard that is applicable for the whole community. That is called the AMC spirit.

That is why I love the AMC. It is not just conference, or seminar, or new standard: it is a movement. It is a way of life. We do good things.

Because we are good, we also contribute in lifting the safety in aviation to whole new level; a level never experienced before. To remind you: because of our hard work and the hard work of many engineers elsewhere, we did not have any catastrophic accidents last year. That is a great achievement. I hope that we can hold that level of safety and fly one more year without incident.

That is not easy. Many smart engineers were running the aviation industry over the last 100 years, but we are the first group who can proudly say that in our lifetime, we had the first year (2017) without commercial airplane accident. That is something to be proud of, knowing that every second of every day, we had about 10,000 aircraft in the air. That is a big number.

Saying that, I would like to ask you a favor. Tell your friends at other airlines to come to the AMC conference because they can learn a lot, and that knowledge will help their airline gain some profit or savings. Simultaneously, we all can learn from them. They have the same challenges we have, but collaboration will help us overcome the challenges.

Once we enter the meeting room and start the open forum discussion, we should speak loudly and express our concerns. The open forum is the place to solve problems.

One piece of advice for the upcoming AMC in Dallas: Do not listen to those who say, "Talk low, talk slow, and don't say too much," because you are the engineer. Cowboys and cowgirls were here before first aircraft was flying. They know a lot about horses and shooting, but not about airplanes. At the AMC, we want you to talk low, talk slow, but say enough to make enable us to fix problems. That is what the open forum is all about. I would also like to remind you that this year, we have great seminars. We will start with the Big Data symposium, which is a joined project with our engineers from AEEC. Tuesday is the symposium about ADS-B, which was requested by many people at the AMC in Milwaukee. Finally, the third seminar block is on Wednesday, and it is about the continuous story of Pitot tubes. We managed to get excellent speakers, the best in the industry, and we are certainly proud of that. Honestly, we are very successful in getting the best speakers in the industry. I have attended many conferences worldwide, but at AMC, I am always impressed with the high quality of experts: they are the best in the industry and are willing and able to share their knowledge.

Finally, in the industry session, I will update you about two important standards we are working on. The first one is the Technical Support Data Package (TSDP) standard, which will define the content of data packages required to build the test setup or TPS. The second one is under the Obsolescence Management Guidance (OMG) Working Group. It is the update of the old obsolescence standard, rewritten and customized for 2018 and coming years.

It will be one more great conference with a variety of subjects to discuss. I will be there and I urge you to come to Dallas too.

One saying is: "Boots, Hats, and Cowboys...Nothing else matters!!"

For the AMC Conference, I would say:

Open Forum, Symposiums, Standards, and Engineers... Nothing else matters!!

See you in Dallas!

AMC Opening Speech Dallas 2018

My fellow engineers! On behalf of the AMC | AEEC, I wish you all the best here at the conference in Dallas. Welcome to Dallas!

My name is J.R.

No, I am not J.R. I am just joking. My name is Marijan Jozic, and I am the Chairman of the AMC conference. Now I am a Texas cowboy, as you can see. First, I would like to remind you that it all started about 115 years ago, when two bicycle makers, known as the Wright Brothers, built the Wright Flyer and did the first flight at Kitty Hawk, North Carolina.

But aviation was a dangerous activity.

The first fatal accident was on September 17, 1908, when Orville Wright was demonstrating that the Flyer could carry passengers. The Flyer crashed and the passenger, Lt. Thomas Selfridge, was the first man killed in an airplane crash.

The Wright brothers flew five years without accident (with less than 100 flights per year).

Now, 100 years later, we have 10,000 aircraft in the air every second of the day, the whole year long. We are flying 100,000 flights per day, which is 37 million flights per year.

Let me give you some numbers: altogether, there are 40 thousand airplanes, including about 30 thousand commercial airplanes, flying. They are flying in 170 countries and using 8,500 airports. A million people are flying on a daily basis, meaning 3.1 billion are flying per year. That is a lot of flying. Think for a second about that number when I tell you the following: 2017 was an extraordinary year. In 2017, there was not one fatal accident with a transport aircraft.

The last fatal accident was November 28, 2016, in Columbia. It is very important to remember that. The latest crash in Moscow was in February 11, 2018, meaning that we had 440 days without any fatal accident. Such a level of safety was not easily achieved. We in this ballroom have contributed to that level of safety. Only we know how much effort is required to fly safe, reliable, and cost effective.

Frankly, engineers are not getting enough credit for that. It was not easy to get this far. Think about every solved issue at the conference, every new standard, every decision we made, every symposium, every success story. Everything contributed to a high level of safety.

Do you think the same that I think? I think: It is because of me! I am sure you also think it is because of you.

And it is, indeed, because of you, my fellow engineers!

Although there is not much publicity about us engineers in the press, I wanted to mention it at the opening of the 69th AMC conference and give you my compliments for such a big achievement. Hats off to my fellow engineers. After saying that, I would like to fire up and open our great conference. As you know, it sounds like this: Bang!

AMC Closing Remarks by Marijan Jozic,

Good morning everybody.

I would like to tell you, today is April 26. It is World Intellectual Property Day. We know everything about Intellectual Property.

Saying that, I know we all are engineers, so we love statistics and numbers. I will give you some updates. We have 761 attendees from all continents except Antarctica. We hope next year we will have some South Pole representatives.
We had, as you know, 228 discussion items.
48 are held open, which is 20%.
25 items were success stories, which is 11%.
There were 13 temperature changes in the ballroom as a result of 2,000 complaints sent to Smitty.
80% of those bags of pretzels are still available here to open, so you can take some of them home.
There were 30 hospitality suites, and the first day, we had 50 new attendees, which is very good.

Our lost and found department informed us that there were 1 tablet lost,
1 wedding ring (I do not know how they managed to lose a wedding ring, so now they have only one ring and it is suffering),
2 laptops,
4 cell phones,
1 Fitbit, and
1 pair of reading glasses.
The Fitbit and reading glasses can be collected at the registration desk. They are found. There is hope.
In our maintenance log, it was recorded that sink number 2 in the men's room was inoperative for 4 days. We could not wash our hands there. Our electric light officer Smitty was asked to change the color of the backstage to Southwest colors, which are now on the stage.

Saying that, I wish you a safe flight home. Thank you for being here and being so cooperative.
Please give an applause to yourselves, and we will hammer the end of the conference.

Over the Poles in 2018

Tickets are available to take part in the Polar Explorer Flight.

LinkedIn is a powerful media tool. I have been a member for a few years. It is very interesting to see the effect of posting to LinkedIn. Of course, it may or may not have an effect. Just last year, I was posting many announcements for the Mechanical Maintenance Conference (MMC) in Cleveland. The MMC was successful and I secretly hope that it was because of me and my LinkedIn posts. Recently, I made a new LinkedIn friend. His name is Roberto Aldana from Polar Explorer Flight. He told a story that had me scratching my head and caused me to do some investigation on the internet.

Seemingly, I started to behave like millennial. Every bit of information is in my tablet or in my smart phone. And all so easily accessible! If you can read you can satisfy your desire for knowledge and information. So my little computer told me that the first commercial Transpolar flight in over 40 years will take place in 2018 and my new friend Mr. Aldana is involved.

Let me first explain what will happen in 2018 and then I will explain its history. The flight plan is as follows:

The flight starts in New York (JFK) on October 26, 2018. From there, in one giant nonstop leap (6,382 miles, 11 hours, 35 minutes) the flight travels southwards to Río Gallegos, at the southernmost tip of Argentina. The flight then travels across the entire continent of Antarctica for a spectacular daytime crossing directly over the geographical South Pole (6,663 miles, 12 hours, 20 minutes) to Perth, Western Australia. The flight then turns north, for a journey to Beijing, China (4,962 miles, 9 hours, 10 minutes). It then crosses continental Russia and the Siberian Plateau over the geographic North Pole to New York (6,710 miles, 12 hours, 45 minutes), landing back at JFK around 2:00 p.m., approximately 50 hours later. Actually, it should be 46 hours and 5 minutes of flying time.

Which brings me to the question: Do you know the Jules Verne story? It was about travelling around the world in 80 days. Well, the Polar Explorer Flight is also around the globe but not in 80 days, nor in 80 hours, but in 50 hours. Despite all of the technical advancements made in space and aerospace during the past half-century, to this day far more people have orbited the earth in space than have made a complete circumnavigation of the world over both poles. How cool is that?

The flight will be executed with A340-500. It is not the fastest airplane, but it can stay in the air for quite a long time. It will stop for refueling at three airports and execute the flight in 46 hours. It is quite an undertaking to organize such a flight as a commercial venue. Of course, the airplane should be well prepared for the journey. In the southern part of the globe, especially in Antarctica, there are no airports and no people. Therefore, the airplane should be double-checked and triple-checked to be close to certain that nothing unexpected can happen. That is what we do every day at our respective airlines, although we do not fly above Antarctica every day. How cool is that!

A Boeing 747SP operating as PanAm flight 50 flew from San Francisco to San Francisco via both poles in 54 hours, 7 minutes, 12 seconds on October 2830, 1977. These days you do not see many B747-SPs around. For the millennials among us, the B747-SP is a very unique B747 that is much shorter than the B747s we are used to seeing. It looks like a B747, feels like B747, and sounds like B747, but it is shorter than one of today's B747s, and it can fly longer the "normal" B747 (6,700 mi/10,800 km). Again, returning to my millennial inquisitive behavior: Wikipedia says that only 45 B747SPs were built, and the idea came from PanAm. Before the B747400, the B747SP was the only airplane that could fly the longest routes. The last one was built in 1989, 29 years ago.

Airbus A340 in Maho Bay, St. Martin

It is fantastic to say that after 41 years, there will be a circumpolar flight again. There is a big chance that it will enter the Guinness World Book of Records because it will beat the 41-year old PanAm record trip.

Looking at the cruise speed, it looks like the story of the turtle and the hare. The cruise speed of an A340 is 871 km/h and the cruise speed of a B747-SP is 914 km/h. It is quite a difference (43 km/h).

If you plan well and choose the best routes, you can fly slow and steady and still finish 8 hours earlier. How cool is that?

I am curious to see how long it will take to break this new record. The Boeing B787 and Airbus A350 can fly very far as well. The cruise speed of the B787 is 901 km/h. It could easily break the record, but it has only 2 engines. I am not sure if it could qualify to fly ETOPS all the way above the global poles? If not, we might wait quite a long time to do this again.

In the meantime, my friend Roberto might have some seats available. It is not inexpensive, but you could enter the history books. Just sell your car and book your ticket for October 26, 2018. And pack some lunches!

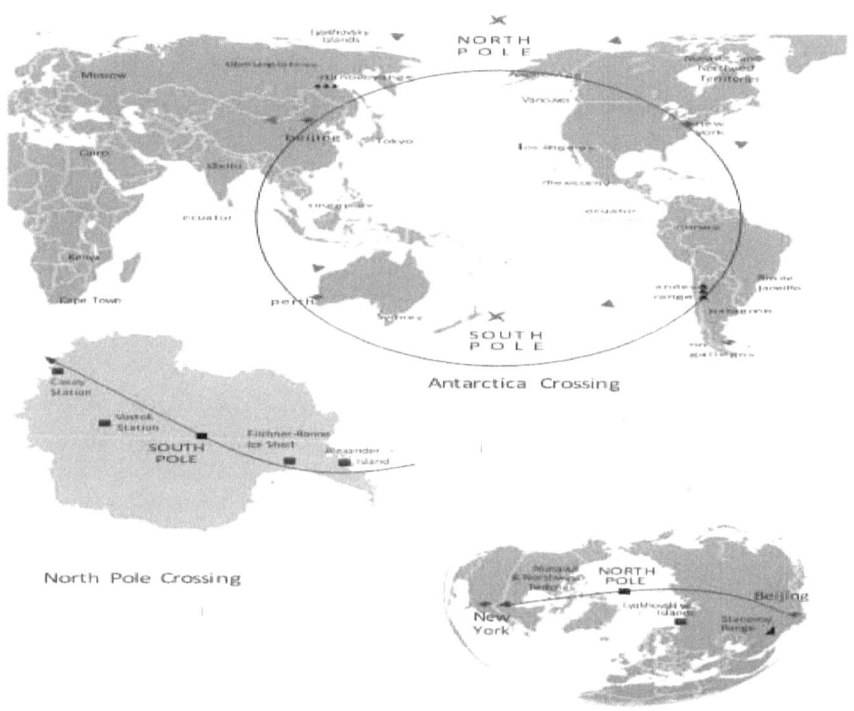

Antarctica Crossing

North Pole Crossing

Prague 2019!
We are ready to Rock and Roll!

Just a few years ago, we had the AMC Conference in Prague, Czech Republic. Prague was one of biggest and most important cities in the Habsburg Monarchy. The Habsburg Monarchy was a dual monarchy better known as the Austrian-Hungarian Monarchy. It has few remarkable cities: Vienna (the capital of the monarchy), Budapest, Prague, Bratislava, and Zagreb. Each of those cities is now a capital of their respective countries: Austria, Czech Republic, Slovakia, Hungary, and Croatia. These cities are still beautiful as jewels, and Prague is a diamond among them. Ask anyone who attended the AMC a few years ago and they will tell you that I am right. But, the city is not the true reason why we are going to Prague. We are going for the AMC Conference. And here is probably the biggest surprise: it will be a bit different than last time. Let me explain.

Last year, we launched the Mechanical Maintenance Conference (MMC) where we had the opportunity to work together with aircraft mechanical engineers. We solved problems in open forum sessions and we educated the crowd in our symposiums. The conference was just two days and we managed to demonstrate that every single minute of the MMC was valuable. Our mechanical friends were absolutely delighted, and the first MMC Conference was an ultimate success. This past April, the AMC Conference was held in Dallas, Texas. Figuratively and literally, there was a big storm above the AMC Steering Committee meetings during the AMC. You may not have noticed that the AMC Steering Committee had a rough ride in Dallas. We deliberated, talked, discussed, checked, and rechecked future schedules and strategies. We discussed issues with the AEEC Executive Committee as well as the SAE ITC leadership team and eventually cut the Gordian knot.

Here is the result. The MMC Conference will be collocated with AMC Conference and the AEEC General Session. Yes, I will repeat: in Prague, we will have a triple venue: AMC-MMC-AEEC. Three conferences in one location. We are convinced that this will be the most cost-effective conference in the history of aviation. With these three events, we will be covering the entirety of all aircraft systems—well, except primary structures and powerplants. Regardless, the avionics wiring keeps airplanes flying. Everyone knows that! In other words, an aircraft is just a bunch of parts flying in close formation held together by wiring. Now you know the secret. Unfortunately, creating a three-sided event is not easy. There are many people involved to be able to compress so much into the fastest four days in aviation. Everyone is dedicated to the events' success: SAE ITC, ARINC IA, AMC, and AEEC Leadership. Even the hotel leadership! On our list of goals and tasks: establish new guidelines regarding questions, select symposiums and seminars, plan to fit it all in the program, coordinate with AEEC, communicate with OEMs, choose moderators, advertise, etc. Of course, we will need to deal with last minute changes and everybody will be asked to cooperate, but we have no doubt that we can do it. In October, there was an AMC Steering Committee meeting in Amsterdam where we will laid out the plan and started to finalize some of the details.

The last bits and pieces will be decided at the AMC Steering Committee winter meeting, and we will then be ready to rock and roll. I am so proud and yet humbled to be the AMC Chairman as we go into its 70th year. I am in awe of all the people that participated in the AMC, shaping aviation history.

In all those years, the AMC has adapted to industry changes. The first 50 years were the years of amazing aviation industry growth. After that, we saw years of deregulation, merging of airlines, merging of OEMs, and in the last couple of years, we have seen the Intellectual Property wars. Through all of these, the AMC managed to adapt. It was not always easy; sometimes it was hard. The aviation industry was changing but gaining experience. The types of AMC questions changed but we anticipated that. The classic standardization of radios was replaced by standardizations of procedures and processes. We have educated the AMC about intellectual property, obsolescence, PMA parts, 3D printing, statistics, Big Data, predictive maintenance, ADS-B, and many other things. We can proudly say that we have made decisions that must be followed by others, not only the AMC participants. They had a chance to join us, but they decided not to. The AMC took the lead and they had no other choice than to follow what we decided. That is the power of the AMC and ARINC Industry Activities. Finally, all avionics people are one big family. We know each other, we collaborate, have the greatest network in aviation, and we are really good. In Prague, our family will grow further. Mechanical engineers will be added to an already strong group of avionics engineers. Again, the AMC conference has adapted. Ten years ago, AMC Chairman Mitch Klink from FedEx announced that there was a new definition for avionics components: Avionics is every component with a wire attached. People were thinking that such a definition was too trivial. But look at the Boeing B787 and Airbus A350: every LRU in every ATA series has at least one wire attached. There is not a valve, check valve, actuator, blower, etc., without at least a sensor. And any sensor needs a wire to transfer the data to other device.

And there is software, almost everywhere. The B787 is generating 500 GB of data after every flight. We are using maybe 5%. The rest is deleted forever. The last chapter in this history book has not been written. The intellectual property wars will soon start about the ownership of data. If it is my aircraft, it is my data? Why should I give it out for free? Those questions are not yet answered, but we know that if you control the data, you also control the materials. The AMC and MMC must contribute to these discussions and educate others.   Some new developments like block chain are normal in other industries. Before we know it, block chain will become the new headache in aviation. We should take care not to be late in our understanding. The AMC's ambition is to lead because we know that leading is better than following. But leading also has the price. We need now more than ever smart young people to wrestle with these new technologies and developments.

Therefore, please urge everyone to attend our conferences. At this moment, we have the whole world in the palm of our hand. We must keep that status, but at the same time, we must challenge status quo and keep adapting to every new development: new technologies, new aircraft, a new concept of maintenance, or a new format of conference. So let's rock and roll! Prague!

How I got my AMC in Prague

Just one memo was required with the following content:

Subject: KLM to host AMC (MMC/AEEC) conference

Everything what we do at the AMC conference is constantly challenging the status quo. We are always thinking differently and strive for perfection. We are not satisfied with the present situation and we always want something different and better. That is why we are the Industry Activity.

The three pylons to obtain perfection and reach our desire are Aviation Standards, Aviation Conferences and Seminars. We mobilize airlines, OEM's and airframers to join forces and improve the aviation industry.

Aviation Standards are developed to standardize designs of aircraft and new systems, shop procedures and test equipment. Maintenance Conferences like AMC (Avionics Maintenance Conference), MMC (Mechanical Maintenance Conference), and FSEMC (Flight simulator Engineering and Maintenance Conference) are established to fix problems in open forum discussions and collaboration. Finally, the seminars are here to educate our engineers and provide the solid knowledge base about the newest developments in aviation.

Besides that all at our conferences, you get in contact with 150 suppliers and 4 airframers and 35 airlines at the same spot. We can exchange experience with 700 or more top engineers. That is how we run the modern aviation. KLM is a prominent member of the Aviation community and was the chairman of the AMC for the last 7 years and the MMC for two years.

Do you want to host the next AMC/MMC conference in Prague (April 2019) and expose the KLM brand to 700 professionals for 4 days and show them why KLM is the leading MRO?

And the answer from my VP was: **YES**! Go for it!

Welcome to Prague!

Welcome note in AMC-MMC Program 2019

We were in Prague with AMC | AEEC in 2015. It was great event. Our purpose was, as always, to improve maintenance and lower costs. For 70 years, we have been organizing conferences and maintaining our purpose, our WHY. All this time, we have been asking ourselves if there was another way to do the maintenance or repair more efficiently and cost effectively. That is why we have been seeking solutions for 70 years. And we are successful! We help save millions of dollars for our airline members. We set standards not only for our members, but for the whole world. Finally, we educate. We are very successful in organizing high-quality seminars and attracting excellent speakers, the best in the industry.

Prague 2019 will be special. Although we have practiced our purpose for 70 years, this time, we have a few new and exciting changes. Let me tell you about them.

Prague 2019 will be bigger than ever. We will have one more event called the Mechanical Maintenance Conference (MMC). As you are aware, the first MMC was in Cleveland, Ohio. The AMC Steering Committee organized it for our fellow mechanical engineers. The first inaugural conference was a success. This year, we decided to hold it together with AMC and AEEC. Therefore, the full name is AEEC | AMC & MMC conference. We extended the ATA range and open forum for all aviation questions with exception of power plant and primary structure. We expect great cooperation and fantastic open forum discussions. Wait and see!

There is also an organizational change. Airline Avionics Institute (AAI), who was our partner for 50 years, will hand over duties to ARINC IA. This means ARINC IA staff will have additional responsibilities, like organizing coffee brakes, lunches, etc. Besides that, everything regarding Volare awards was also handed over to ARINC IA. To guarantee that Volare award recipients will be selected honestly and unbiasedly, the award selection will be handled by a separate committee totally independent of the organization. Wait and see!

In Prague, we will do an experiment with attendees. It will be something interactive and very special. To keep it a surprise, I will not tell you much about it. Wait and see!   Last time, our attendees announced that the AMC conference in Prague 2015 had the best food ever experienced at a conference. We hope that this time we can prove that it can be even better. Wait and see!

This is the seventh time that I am writing the Welcome in the AMC & MMC Program. I have attended 20 AMC conferences, a few AEEC conferences, the inaugural MMC conference, and many working groups. It was not always easy. The last seven years as a chairman was a wonderful time and I enjoyed every little bit of it. But it is time to hand over the torch to a new chairman, who will be elected after the conference. Somebody told me that the best time to stop is when it hurts a bit. Well, it looks like I have chosen the right time. I am happy that I can hand over the torch to a new chairman. At the same time, it hurts a bit because I will miss my leading role. The following words are the best description of my feelings:

We'll meet again Don't know where Don't know when But I know we'll meet again some sunny day!

But first and foremost, let's celebrate in Prague the 70th AMC conference (together with MMC and AEEC) and make the best of it, as we always do.

The Chairman

The Last Chairman's Speech

(AMC-MMC-AEEC opening in Prague in front of 700 engineers)

Welcome to Prague,
Welcome to AMC- MMC- AEEC conference.

Prague is one of famous Habsburg cities together with Vienna, Budapest, and Zagreb. We are just starting the 70th AMC conference. That is what I call tradition. In 1949 the AMC conference was established, and our purpose was defined. Our WHY was defined!

Why means purpose—Our purpose!

Our purpose is to constantly challenge the status quo. Always investigating is there another way to do maintenance and repairs. Our purpose is to lower the costs and increase the reliability of our Aircraft Components and the airplane.

Engineers from our first conference and engineers today still see that AMC is actively improving the aviation industry. That is going on for 70 years and our purpose is still the same?

But, How we do that?

The Australians would say: There is more than one way to skin the cat. Therefore, our way to skin the cat is to fight in the following arenas:

- Conferences,
- Seminars
- and Setting new Standards.

The idea was to use those three activities as pylons which support our WHY, our purpose.

The AMC was initiated in 1949 with only open forum discussions because we believed that it was right time to mobilize the world and fix problems. Later we organized working groups to develop Standards and implemented seminars to talk about ideas worth spreading.

What we do is clearer than ever. So, if somebody ask what we do, the answer will be easy and every one of us knows it.

- Fix technical problems in open forum discussions,
- Define standards in our working groups
- We educate our attendees in numerous seminars.

The most important thing of all is to realize that after 70 years, our activity, our purpose, remains the same:

- It is not what we do, it is why we do it!
- It is not what we do, it is why we do it.

Only if we all believe in our WHY (our purpose), we can be successful and attract others to join us.

At the end of the road, all of those outside AMC and ARINC Industry Activities will believe in what we believe, and that is why we are here.

So never forget our purpose which is to reduce operating and life cycle costs of air transport avionics and increase reliability of our components and keep challenging the status quo.

This is my last AMC conference as a chairman. Over the past 7 years, I have created an atmosphere of cooperation, joint effort, and humor to serve our purpose.
You must know the famous words:
It is my fault.
I am sorry.
I will fix it.
I will provide the warranty.
And it is free!

I enjoyed every little bit of it!

The next chairman will be here next year but our purpose, our **why**, will stay unchanged because we believe in it.

I see that you are still with me and that is always a good sign.

So, before applause let's officially open the 70th AMC conference and our collocated MMC and AEEC conferences.

Bang

AMC
Together "We Set the Standard."

AVIONICS MAINTENANCE CONFERENCE

EST. 1949

CHAIRMEN

1949-1950 C. I. Rice Northwest	1972-1974 Andrew Flood Seaboard	1987-1989 Pete Clarke USAir	2005-2009 Axel Mueller Lufthansa
1950-1955 Walter D. Rollick Piedmont	1974-1976 Lauren Nelson Eastern	1989-1991 Pat Windham Delta	2009 - 2012 Mitch Klink FedEx
1955-1959 J. E. M.(Mel) Lagasse Trans-Canada	1976-1977 Andrew Flood Seaboard	1991-1994 Dale Johnson American	2012 - Marijan Jozic KLM
1959-1963 John A. Lehman United	1977-1978 Dean Haney Continental	1994-1995 Pat Windham Delta	
1963-1965 Lee Reilly American	1978-1979 Robert M. Bennett Flying Tigers	1995-1998 Ed Sawyer FedEx	
1965-1967 Peter Sammon Pan American	1979-1981 Colonel Lester Delta	1998-1999 Martin Story Delta	
1967-1969 Henry Harrison Eastern	1981-1984 John J. Laviolette Air Canada	1999-2000 Peter Foessinger Lufthansa	
1969-1971 Merwyn Weaver National	1984-1986 E. R. (Bud) Scanlon Delta	2000-2001 Paul Winkels Lufthansa	
1971-1972 Robert C. Kurtz American	1986-1987 Dennis Logan Pan American	2002-2003 Martin Story Delta	

www.ingramcontent.com/pod-product-compliance
Lightning Source LLC
Chambersburg PA
CBHW030928180526
45163CB00002B/493